The Advances publish reviews and critical articles covering the entire field of normal anatomy (cytology, histology, cyto- and histochemistry, electron microscopy, macroscopy, experimental morphology and embryology and comparative anatomy). Papers dealing with anthropology and clinical morphology will also be accepted with the aim of encouraging co-operation between anatomy and related disciplines.

Papers, which may be in English, French or German, are normally commissioned, but original papers and communications may be submitted and will be considered so long as they deal with a subject comprehensively and meet the requirements of the "Advances".

For speed of publication and breadth of distribution, this journal appears in single issues which can be purchased separately; 6 issues constitute one volume.

It is a fundamental condition that submitted manuscripts have not been, and will not simultaneously be submitted or published elsewhere. With the acceptance of a manuscript for publication, the publisher acquire full and exclusive copyright for all languages and countries. 25 copies of each paper are supplied free of charge.

Die Ergebnisse dienen der Veröffentlichung zusammenfassender und kritischer Artikel aus dem Gesamtgebiet der normalen Anatomie (Cytologie, Histologie, Cyto- und Histochemie, Elektronenmikroskopie, Makroskopie, experimentelle Morphologie und Embryologie und vergleichende Anatomie). Aufgenommen werden ferner Arbeiten anthropologischen und morphologisch-klinischen Inhalts, mit dem Ziel, die Zusammenarbeit zwischen Anatomie und Nachbardisziplinen zu fördern.

Zur Veröffentlichung gelangen in erster Linie angeforderte Manuskripte, jedoch werden auch eingesandte Arbeiten und Originalmitteilungen berücksichtigt, sofern sie ein Gebiet umfassend abhandeln und den Anforderungen der „Ergebnisse" genügen. Die Veröffentlichungen erfolgen in englischer, deutscher und französischer Sprache.

Die Arbeiten erscheinen im Interesse einer raschen Veröffentlichung und einer weiten Verbreitung als einzeln berechnete Hefte; je 6 Hefte bilden einen Band.

Grundsätzlich dürfen nur Arbeiten eingesandt werden, die nicht gleichzeitig an anderer Stelle zur Veröffentlichung eingereicht oder bereits veröffentlicht worden sind. Der Autor verpflichtet sich, seinen Beitrag auch nachträglich nicht an anderer Stelle zu publizieren. Die Mitarbeiter erhalten von ihren Arbeiten zusammen 25 Freiexemplare.

Les résultats publient des sommaires et des articles critiques concernant l'ensemble du domaine de l'anatomie normale (cytologie, histologie, cyto- et histochimie, microscopie électronique, macroscopie, morphologie expérimentale, embryologie et anatomie comparée). Seront publiés en outre les articles traitant de l'anthropologie et de la morphologie clinique, en vue d'encourager la collaboration entre l'anatomie et les disciplines voisines.

Seront publiés en priorité les articles expressément demandés, nous tiendrons toutefois compte des articles qui nous seront envoyés dans la mesure où ils traitent d'un sejet dans son ensemble et correspondent aux standards des «Revues». Les publications seront faites en langues anglaise, allemande et française.

Dans l'intérêt d'une publication rapide et d'une large diffusion les travaux publiés paraitront dans des cahiers individuels, diffusés séparément: 6 cahiers forment un volume.

En principe, seuls les manuscrits qui n'ont encore été publiés ni dans le pays d'origine ni à l'éntranger peuvent nous être soumis. L'auteur s'engage en outre à ne pas les publier ailleurs ultérieurement. Les auteurs recevront 25 exemplaires gratuits de leur publication.

Manuscripts should be addressed to/Manuskripte sind zu senden an/Envoyer les manuscrits à:

Prof. Dr. A. BRODAL, Universitetet i Oslo, Anatomisk Institutt, Karl Johans Gate 47 (Domus Media), Oslo 1/Norwegen

Prof. W. HILD, Department of Anatomy, Medical Branch, The University of Texas, Galveston, Texas 77550/USA

Prof. Dr. J. van LIMBORGH, Universiteit van Amsterdam, Anatomisch-Embryologisch Laboratorium, Mauritskade 61, Amsterdam-O/Holland

Prof. Dr. R. ORTMANN, Anatomisches Institut der Universität, Lindenburg, D-5000 Köln-Lindenthal

Prof. Dr. T. H. SCHIEBLER, Anatomisches Institut der Universität, Koellikerstraße 6, D-8700 Würzburg

Prof. Dr. G. TÖNDURY, Direktion der Anatomie, Gloriastraße 19, CH-8006 Zürich/Schweiz

Prof. Dr. E. WOLFF, Collège de France, Laboratoire d'Embryologie Expérimentale, 49 Avenue de la belle Gabrielle, Nogent-sur-Marne 94/Frankreich

Advances in Anatomy, Embryology and Cell Biology
Ergebnisse der Anatomie und Entwicklungsgeschichte
Revues d'anatomie et de morphologie expérimentale

Vol. 53 · Fasc 4

Editors:
A. Brodal, Oslo · W. Hild, Galveston · J. van Limborgh,
Amsterdam · R. Ortmann, Köln · T. H. Schiebler, Würzburg
G. Töndury, Zürich · E. Wolff, Paris

Brian Keith Hall

Chondrogenesis
of the Somitic Mesoderm

With 5 Figures

Springer-Verlag Berlin Heidelberg New York 1977

Professor Dr. Brian K. Hall, Department of Biology, Life Sciences Centre, Dalhousie University, Halifax, Nova Scotia, Canada B3H 4J1

ISBN-13: 978-3-540-08464-8   e-ISBN-13: 978-3-642-66766-4
DOI: 10.1007/ 978-3-642-66766-4

Library of Congress Cataloging in Publication Data. Hall, Brian K. 1941–   . Chondrogenesis of the somitic mesoderm. (Advances in anatomy, embryology, and cell biology; 53/4) Bibliography: p.   Includes index. 1. Chondrogenesis. 2. Mesoderm. 3. Somite. I. Title. II. Series. QL801.E67 vol. 53/4 [QL979.5] 574.4'08s [596'.01'7] 77-21183

This work is subject to copyright. All rights are reserved, whether the whole or part of the materials is concerned specifically those of translation, reprinting, re-use of illustrations, broadcasting, reproduction by photocopying machine or similar means, and storage in data banks.

Under § 54 of the German Copyright Law where copies are made for other than private use, a fee is payable to the publisher, the amount of the fee to be determined by agreement with the publisher.
© Springer-Verlag Berlin-Heidelberg 1977.

The use of registered names, trademarks, etc. in this publication does not imply, even in the absence of a specific statement, that such names are exempt from the relevant protective laws and regulations and therefore free for general use.

Composition: H. Sturtz AG, Universitatsdruckerei, Wurzburg
2121/3321-543210

Contents

Introduction 7

1. Development and Division of the Somite 8
    1.1. Migration of Pre-somitic Mesoderm 8
    1.2. Segmentation of the Mesoderm into Somites 8
    1.3. Subdivision of the Somites 9
        1.3.1. Sclerotome, Dermatome, Myotome 9
        1.3.2. Resegmentation of the Sclerotome 10
    1.4. Spinal Ganglia 11

2. Self Differentiation or Induction 12
    2.1. 1925–1940: Grafts to the Avian Chorioallantoic Membrane 13
    2.2. 1940's–1950's: In vivo Transplantation and Extirpation 14
        2.2.1. Amphibia 14
        2.2.2. Pisces 15
        2.2.3. Aves 15
    2.3. 1950's Onwards'': In vitro Chondrogenesis 16

3. The Inductive Tissues 21
    3.1. Integrity of the Inducer and Vertebral Morphogenesis 21
    3.2. When Do the Inducers Lose Their Inductive Ability? 22
    3.3. Ability of Dermomyotome to Chondrify 23
    3.4. The Search for a Pure Inducer 24

4. Ectopic Inducers 25
    4.1. Can Ectoderm Act Inductively on Somites? 25
    4.2. Other Ectopic (heterologous) Inducers 28

5. Synthesis of Glycosaminoglycans and Onset of Cartilage Differentiation 30

6. Production of Extracellular Matrix by Notochord and by Spinal Cord 31
    6.1. Glycosaminoglycans 31
    6.2. Collagen 32

7. Function of Extracellular Matrix Produced by Notochord and Spinal Cord .. 36

8. Discussion and Conclusions 38

9. Summary 41

References 42

Subject Index 48

Author Index 50

Introduction

Chondrification of the somitic mesoderm producing the primordia of the vertebrae has been a favourite topic of investigation by those interested in understanding the mechanisms underlying development and differentiation. Study of this tissue involves the investigation of cell proliferation, migration and cytodifferentiation, synthesis of extra cellular matricial products, inductive tissue interactions, and tissue morphogenesis — some of the major unsolved problems of modern developmental biology. The topic is a large one and various aspects have been reviewed in the past, notably by Holtzer (1959; 1961), Lash (1963a), Strudel (1967), Lash (1968a, c), Holtzer (1968), Holtzer and Abbott (1968), Holtzer and Mayne (1973) and Levitt and Dorfman (1974). Feeling that the time was appropriate for a overview of the whole field of somitic chondrogenesis, I have prepared the present monograph. It covers the development of experimental studies from the initial grafting of somites to avian chorio-allantoic membranes in the 1920's, through the extirpation experiments of the 1940's, the culturing of the 1950's and the search for pure inducers, to the refocussing of attention on the environment and matrix procucts which is dominating the studies of the 1970's. Many of the studies have utilized the embryonic chick as the experimental animal.

Wherever possible the ages of the embryos used have been given as Hamilton and Hamburger stages as well as number of days of incubation, and in some cases as number of somites as well (see Table 10).

1. Development and Division of the Somite

1.1. Migration of Pre-somitic Mesoderm

In tracing the origin of the vertebrae it is appropriate to begin with the development, segmentation and sequential subdivision of the paraxial mesoderm into pre-somitic mesoderm, pairs of somites, and sclerotome, dermatome and myotome. Williams (1910) provided the first comprehensive account of formation and segmentation of the somites (in the embryonic chick) and recently Lipton and Jacobson (1974a) have provided an elegant account of these early stages.

At HH 2–3 in the embryonic chick, paraxial mesoderm adheres to the basement membrane of the epiblast above and to the cells of the hypoblast below. By HH 4–6 mesodermal cells have multiplied to form a cell layer 4–6 cells thick and the pre-somitic mesoderm may be identified as a line of cells preferentially adhering to the basement membrane of the presumptive neural ectoderm. As the neural plate moves toward the mid line (HH 6–7) the cells of the presumptive somite become more compacted, eliminating extra cellular spaces and forming new cell to cell junctions at their apical ends. They remain attached to the neural plate so that as the neural plate elevates at HH 8–9, the somitic mesoderm moves mediad with it, only then separating from the neural plate (now the neural tube) and from the developing notochord (Lipton and Jacobson 1972a, Figs. 3–9). This separation may be facilitated by the accumulating extracellular matrix which is being produced by both the notochord and by the neural tube (see pp. 31–36). By HH stage 12 bundles of fibrils connect the somite with the notochord and with the neural tube, maintaining a very close association between somitic mesoderm, notochord and neural tube.

1.2. Segmentation of the Mesoderm into Somites

There are at least four major theories available to explain segmentation of the mesoderm into somites in the embryonic chick. These are (a) presence of a somite organizing centre adjacent to the primitive streak; (b) the normal regressive movements of the primitive streak; (c) presence of the notochord; (d) presence of the neural tube. These have been reviewed by Bellairs (1971) and by Deuchar (1975). The weight of the evidence seems to favour b, c, and d and excludes a. The possible role played by the segmentally arranged spinal ganglia in somite segmentation will be discussed on p. 11.

Lipson and Jacobson (1974b) and Packard and Jacobson (1976) have excellent evidence to support their thesis that the neural plate imposes a "segmentation prepattern" on the pre-somitic mesoderm and that this pattern is released during subsequent migration of the mesoderm. Their experimental evidence is that 180° rotation of an area of the epiblast of HH 4–6 embryos which places presumptive neural plate over presumptive lateral plate mesoderm, induces the formation of segmentally arranged somites within the lateral plate mesoderm. They further show that this segmentation prepattern is normally released during the regressive movement of the primitive streak as the notochord shears the mesoderm into lateral halves. An incision alongside the primitive streak will substitute for this movement and evoke simultaneous formation of somites along the axis. Normally the somites do not form simultaneously but in an anteroposterior sequence, just behind the advancing notochord.

1.3. Subdivision of the Somites

1.3.1. Sclerotome, Dermatome, Myotome

The somite when it first forms is solid. It then hollows-out, each somite as it develops having a larger cavity than did the preceeding somite. Each successive somite is larger overall, more advanced when it arises, and undergoes further development more rapidly, than the somite immediately anterior to it.

Each somite subdivides into three major components. The dorsomedial wall of each hollow somite forms the *myotome*, which will subsequently form the somatic musculature. The dorso-lateral edge of each somite forms the *dermatome*, and is the source of the connective tissue of the skin. There is evidence (Langman and Nelson, 1968) which indicates that these two components should be regarded as a *dermomyotome* for cells from the "Dermatome" produce both myoblasts and fibroblasts. The ventromedial edge of the inner wall of each somite forms the *sclerotome* and will form the vertebrae and the intervertebral discs of the vertebral column (Fig. 1).

Whereas the dermomyotome remains compact and elongates mediolaterally, the sclerotome breaks-up into a nest of cells. These cells move to the intra-somitic cavity where they loose their epithelial shape and become "mesenchymal" in appearance. The stimulus for this break-up is unknown. This phase of cell separation, which begins at HH 11 (40–45 hours) in the chick, and during the fourth week in man (Arey, 1974), is followed by a phase of cell migration away from the rest of the somite (during HH 13–18 in the chick). The sheet of cells so produced, takes up a position between the notochord and the ventral portion of the spinal cord (Fig. 1).

These changes do not take place simultaneously in all somites. Minor (1973) in an electron microscopical study of somite formation indicates that at HH 17 (54–60 hours) the cranial somites are organising into dermomyotome and sclerotome whilst the caudal somites are still segmenting, although the latter already have the full complement of cellular organelles.

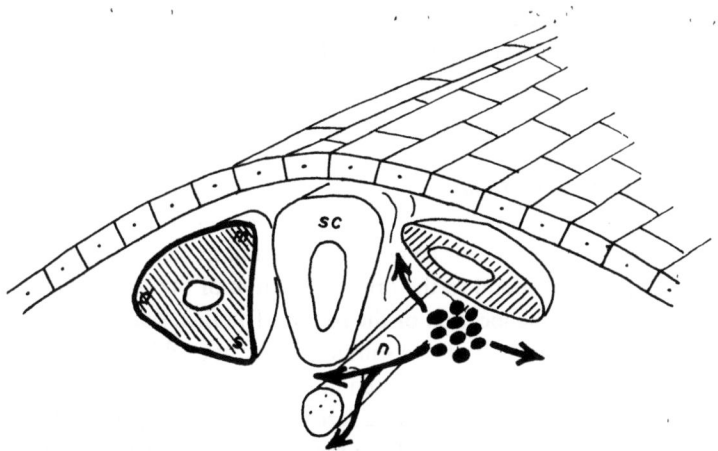

Fig. 1. Transverse section through a vertebrate embryo to show, left of the spinal cord (*sc*) and notochord (*n*), the position of the dermatome (*d*), myotome (*m*) and sclerotome (*s*) before migration of sclerotome. On the right of the spinal cord is shown the direction of migration of sclerotomal cells (black) after subdivision of the somite

At HH 18 in the chick the more cranial sclerotomal cells are migrating around the notochord and the developing spinal cord. Conveniently a space is available lateral to the embryonic axis and between the notochord and the spinal cord. There is also a ventro-lateral migration of cells between the myotomes, delineating the forerunners of the ribs, the costal processes (Arey, 1974).

The origin of the ribs has been the subject of much debate, principally as to whether they are derived from som*a*tic or from som*i*tic mesoderm. The debate seems to have been halted, at least temporarily, in favour of somitic origin. Sweeney and Watterson (1969) blocked the lateral migration of the somitic mesoderm with a physical barrier, and Pinot (1969) blocked migration by X-irradiation. In both cases ribs failed to develop. More recently Chevallier (1975) has used homo- and heterospecific transplantation between chick and quail tissues in further confirmation of a somitic origin for the ribs.

As will be discussed on pp 31–36 the notochord and the ventral portion of the spinal cord both actively produce extracellular matrix *before* the sclerotome cells begin to migrate around them. Therefore the migrating cells (HH 18ff) are moving into an extracellular environment rich in collagen fibres and in glycosaminoglycans (sulphated and unsulphated). Some of this matrix may indeed be integrated into the matrix which later surrounds the sclerotomal cells. By HH 23 matrix is found between those sclerotome cells which are adjacent to the notochord but not in those still more distant from the notochord. Subsequently there is a gradual spreading of extracellular matrix peripherally. Between HH 23–29, the peri-notochordal sclerotome cells in the future vertebral areas consists of a mixture of mesenchymal cells (prechondroblasts) and occasional fibroblasts, whilst in the intervertebral areas fibroblasts predominate (Strudel, pers. comm). By HH 34 the prechondroblasts have become chondroblasts, and by HH 37 they are recognisable chondrocytes (Olson and Low, 1971). Crissman and Low (1974) have described the electron microscopy of subsequent development of the vertebrae over HH 35–42 ($8\frac{1}{2}-16$ days). Within this period chondrocytes complete their phase of deposition and begin to undergo necrosis as a prelude to endochondral replacement by bone. There is considerable deposition of osteoid on uncalcified cartilage matrix. (In long bones osteoid is deposited onto calcified cartilage matrix.)

1. 3. 2. Resegmentation of the Sclerotome

The final position that the sclerotomal cells come to occupy and in which the vertebrae finally develop does not correspond to the original position of the somite but is "intersegmental" with respect to the location of the somite. The final sclerotome corresponds to the combined caudal half of one somite and the cranial half of the next most caudal somite. This concept was first presented in a unified way by Remak (1855). Each half of the sclerotome is known as a sclerotomite (after Piiper, 1928). Thus the atlas, the first cervical vertebra, arises from the halves of the fifth and sixth permanent somites. This arrangement is achieved through the cephalad migration of cells from the cephalic portion of one sclerotome and the caudad migration of cells from the caudal half of the next (Patten and Carlson, 1974, Figs. 16–11).

Between the vertebrae, mesenchymal cells surround the notochord to form the intervertebral discs. This mesenchyme comes from the dense caudal half of each sclerotomite, the cells migrating cephalad to form the disc. The notochord within the disc forms the nucleus pulposus. If a remnant of the notochord persists within the vertebral column it can become tumorous (so called chordomas). The intervertebral disc comes to lie between two adjacent vertebrae and opposite the myotome (i. e. in the position of the original somite). This arrangement allows muscle insertions across the joint and lateral flexure of the vertebral column.

Each lateral sclerotomite may provide just one cell lineage as its contribution to the vertebra. This concept of clonal origins of somite derivatives is based on the work of Gearhart and Mintz (1972) and of Moore and Mintz (1972) and has been reviewed by Mintz (1971, 1972). Embryos derived from fusion of blastomeres (usually 8-cell stage mice) are used to follow strain-specific characteristics of the vertebrae. These tetraparental mice indicate that each of the two caudal and two cranial sclerotomites may provide one cell lineage to the vertebra and thus that the thirty vertebrae could derive from just 120 cell lineages (clones). At the present time such lineages cannot precisely pinpoint the time of determination of each line (see Abbott et al., 1972 for critical comment on the method).

1. 4. Spinal Ganglia

The cartilaginous rudiments of the vertebrae are not continuous around the notochord and spinal cord but are interrupted by the spinal ganglia. The latter have developed from neural crest cells which migrated from the neural folds to the median face of each somite and they are present as segmental structures before the vertebrae develop (Strudel, 1967). Therefore it is possible that they act as mechanical barriers to the migration of sclerotome cells, reflecting them cephalad and caudal to effect the resegmentation of the sclerotome. Indeed the ganglia may actively repulse the approaching mesenchymal cells (Holtzer, 1952a; Balinsky, 1975). Feller and Sternberg (1934) from their study of sectioned human embryos, suggested that the spinal ganglia had an involvement in the morphogenesis of the vertebral column. They proposed that the normal morphology of the neural arch (the dorsal component of the vertebra) was dependent on influences from the central nervous system and from the spinal ganglia, whereas central morphology was influenced by the notochord. Subsequent experimental studies have substantiated this view.

Excision of the neural crest in either amphibia (Detwiler, 1934, 1937; Detwiler and Van Dyke, 1934) or in birds (Strudel, 1953a, 1967) leads to reduction or absence of spinal ganglia and to formation of neural arches which fail to segment (Fig. 2). The centra are normally segmented, indicating that the ganglia influence the dorsal but not the ventral sclerotome. Further support comes from grafting avian somites to the chorioallantoic membrane of host embryos, where segmented neural arches only form if spinal ganglia are grafted adjacent to the somites (Williams, 1942). In the absence of spinal ganglia and spinal roots the segmentation of the centra is only normal if the ventral spinal roots are present (Strudel, pers. comm., based on his unpublished data).

Fig. 2. (a) Axial structures from a normal embryo to show the segmental arrangement of the neural arches and centra (stippled), and of the spinal ganglia (black) around the central spinal cord and notochord. (b) The effect of surgical removal of the spinal ganglia. Neural arches develop but are unsegmented. The centra are normal. (c) The effect of surgical removal of the notochord. Centra develop but are unsegmented. The neural arches are normal. (d) The effects of surgical removal of both spinal ganglia are notochord. Unsegmented centra and neural arches result in formation of a solid vertebral column.

2. Self Differentiation or Induction

Is the differentiation of the cartilaginous primordia of the vertebrae an example of self differentiation of the somitic mesoderm or is it dependent upon influences emanating from the adjacent tissues — the notochord, spinal cord, spinal ganglia, spinal nerves, epidermal ectoderm and endoderm? The answer to this question has not remained the same over the course of experimentation of somitic mesoderm and this system is an excellent example of the way in which the current experimental method influences interpretation of developmental processes. Before reviewing the history of this question in detail it will be useful to briefly outline the highlights of the story.

The first experimentation on differentiation of the somitic mesoderm was carried out in 1925 and utilized the then newly-developed method of grafting tissues to the vascularised chorioallantoic membrane of the embryonic chick. This method dominated the studies until the early 1940's. Cartilage was observed to develop in somites grafted without the notochord or spinal cord, with some authors maintaining that these latter structures enhanced chondrogenesis within the grafted somites. Avian embryos were the source of the experimental material.

Then avian embryos were abandoned for a period and their place on the operating table taken by members of the Urodele Amphibia. Horstadius (1944) introduced extirpation procedures which enabled removal of the notochord or spinal cord from early embryos with subsequent disruption of somitic chondrogenesis. Holtzer and Detwiler, working on amphibian embryos and Strudel and Watterson on avian embryos dominate the literature of the early 1950's. During this period it became well established that, in vivo, the spinal cord and the notochord influenced both cartilage differentiation and vertebral morphogenesis.

Then the mammals enter the scene. Grobstein and Holtzer (1955) utilized somites from foetal mice to carry out the first in vitro analysis combining somitic mesoderm with spinal cord. This ushered in an explosive outburst of studies on in vitro chondrogenesis of somitic mesoderm (the 1950's) and established that cartilage would only form from the somitic mesoderm if an inducer were present. Induction was regarded as a rigidly controlled switching on of new developmental events within the somite.

The 1960's ushered in a further switch in interpretation as the observation of Avery, Chow and Holtzer (1956) that cartilage could form from somites in vitro and on the CAM *without* the inducer being present, was followed up. A similar observation for somites grafted to the CAM had been made twenty three years before (Murray and Selby, 1933), but as with so many developmental phenomena, it had to be rediscovered again and again. The new observations in vitro brought new interpretations with attention focussing on the somite as the responding tissue rather than on the inducer as a provider of developmental information. Concepts of induction changed from "instructive information" to "permissive induction". Induction was now regarded as enhancing pre-existing potentials and not as imposing new potentials onto responding tissues. New editions of embryology textbooks were published.

The 1970's see attention returning to the inducer, especially to the role of the extracellular matrix of notochord and spinal cord in enhancement of somitic chondrogenesis. With this outline as background we shall now examine the studies in detail.

2.1. 1925–1940: Grafts to the Avian Chorioallantoic Membrane

During the early 1920's techniques for maintaining tissues outside the body were being developed. Techniques such as organ culture and grafting to the chorioallantoic membrane were being used to test the ability of organ primordia (predominantly avian) to self-differentiate and to regulate. Hoadley seems to have been the first to attempt (or at least to publish his attempts) isolation of somitic mesoderm. He (1925) separated strips of somites from the axes of $1\frac{1}{2}$-day-old chick embryos (HH 10) and grafted them to the CAM of 8 and 9 day old embryos. After 7 to 10 days as grafts, cartilage appeared within the somites, irrespective of whether the spinal cord was included in the graft or not. The grafted somites were not entirely clean but were grafted along with the overlying epidermal ectoderm and perhaps also with the underlying endoderm. The ectoderm will reappear again and again during this story but its inductive role (if it has one) still remains unclear.

A slightly later study, that of Murray and Selby (1933) even trough it predated that later interest in the inductive role of the spinal cord, is in retrospect, extremely illuminating on this point. Somites, either singly or as strips of three (in both cases probably along with the overlying ectoderm) were taken from 2-day-old embryos (HH 12–13) and grafted to the CAM of 8-day-old hosts. Five percent of the single somites and 8 % of the trios produced cartilaginous nodules. (It is of interest, in the light of the future studies on critical mass, that a single somite produced cartilage). Bone and muscle also formed in these grafted somites. If the spinal cord was included with the single somites, the incidence of cartilage increased to 60 %. That the low incidence of cartilage obtained in somites grafted without the axial structure was due to influences from contaminating spinal cord or notochord seems unlikely in the light of the fact that unsegmented paraxial mesoderm from younger embryos produced cartilage in 30 % of the grafts. Therefore paraxial mesoderm has a chondrogenic bias before segmentation into somites occurs, before separation of the sclerotome from the remainder of the somite and before migration of the sclerotomal cells toward the spinal cord. It was to be many years before this fact was rediscovered.

Williams (1942) also grafted groups of three somites to the CAM. He grafted somites from 40 hours (HH 11?) embryos, either with or without the notochord. Frequency of cartilage was greatest when the notochord was included in the graft. Williams concluded that the notochord served a mechanical role and specifically denied it any inductive properties.

The question of the dependence of the somites of the avian embryo on the spinal cord or notochord remained a debated one' into the fifties. Straus and Rawles (1953) grafted somites from embryos with 25 to 28 pairs of somites (HH 16) intracoelomically and observed both cartilage and muscle formation. Again ectoderm may have been included in the grafts.

In summary these early studies indicated that cartilage would form in the absence of the spinal cord or notochord; that more cartilage would form in more grafts if spinal cord or notochord were included in the graft; that cartilage formed in somites isolated from embryos as young HH 10, and from as few as one somite; and that paraxial unsegmented mesoderm could form considerable amounts of cartilage when grafted to CAM.

2.2. 1940's–1950's: In vivo Transplantation and Extirpation

An experiment carried out in 1944 by Horstadius ushered in an era of active investigation on the possible inductive role of the spinal cord and notochord acting on somites in vivo. This initial experiment was performed on an amphibian.

2.2.1. Amphibia

Horstadius (1944) removed the roof of the archenteron (presumptive chorda-mesoderm) from early embryos of *Ambystoma punctatum* and found that, although some cartilage differentiated within the sclerotome, its morphogenesis and regional organisation was severly deranged. Kitchin (1949) removed the notochord from neural-plate stage embryos of *A. mexicanum* and found that vertebral cartilage developed as a fused rod, lacking neural arches. Thus these pioneering studies of the 1940's indicated that notochord was essential for morphogenesis but not for differentiation of vertebral cartilage in these two species of *Ambystoma*.

Holtzer (1951) rotated the spinal cord in embryonic urodeles and found that the axial skeleton which subsequently developed was also inverted with respect to its normal dorso-ventral axis. Not only was the regionalisation of the vertebral column dependent on the notochord, but the spinal cord, and evidently only the ventral spinal cord, also played a role. Holtzer continued his studies (1952a, b) and in a careful morphogenetic study found that manipulation of the size of the neural tube resulted in changes in the size of the vertebral column. Using several species (*A. punctatum, A. tigrinum, Triturus torosus*) he found that removal of the notochord from the embryos resulted in formation of a massive cartilaginous rod ventral to the spinal cord, whereas elevation of the spinal cord to a position dorsal to the somite, led to production of vertebral arches around the spinal cord. He concluded (1952b) that the role of the notochord was secondary to that of the spinal cord in these species. Subsequent study by Holtzer and Detwiler (1953) substantiated this position. Removal of the notochord produced unsegmented vertebrae whereas removal of the neural tube from stage 23—24 embryos left the somites unable to chondrify. Whether the lateral dermomyotome of the somite could participate in cartilage formation was also investigated and will be discussed below (p. 23).

A variety of other experimental approaches used by Detwiler and Holtzer (1956) added to this position. Supernumerary somites grafted ventrolateral to the host somites failed to produce cartilage or muscle, presumably because they were isolated from the axial spinal cord and notochord by the intervening somites of the host. However if the spinal cord was split longitudinally and the somites grafted into the split, they differentiated both into cartilage and into muscle. Holtfreter (1968) obtained cartilage only if axial organs were included with his cultured somites. Thus in *Amblystoma* and in *Triturus* the spinal cord is a prerequisite for sclerotome chondrification and the notochord a prerequisite for cartilage segmentation. Here the studies on the amphibia rests for in the mid 1950's attention turned toward the avian embryo as a much more accessible specimen for studies on embryonic induction.

One study using anuran amphibians has been made Smithberg (1954) using *Rana pipiens*. He found that transplantation of tail somites into the abdomen of developing neurulae, whilst allowing normal muscle development, did not enable cartilage to form.

2.2.2. Pisces

Watterson (1952) removed the neural tube from *Fundulus heteroclitus* at the 4—14 somite stage and observed a subsequent inability of the somites to chondrify. He also made the interesting observation that even when the neural tube was badly disorganised, as long as it was present, a normal neural arch developed, implying that no morphogenetic information was transmitted from the neural tube to the somite. This is in obvious contrast with the situation in the urodele amphibia where integrity of the spinal cord is essential for vertebral morphogenesis and regionalization.

2.2.3. Aves

A very considerable amount of experimental work has been carried out on the embryonic chick, involving extirpation, transplantation, grafting and maintenance in vitro.

Strudel in Paris has been one of the most active workers in the field. Initially (1953a) he removed the neural tube from early embryos and found that, later in development, no neural arches developed. The spinal ganglia did develop, indicating that the neural crest cells were able to differentiate into ganglia in the absence of the neural tube. Removal of the notochord resulted in absence of the body of the vertebra (the centrum) but had no effect on development of the neural arches or the ribs (Fig. 2). Therefore he postulated that the notochord induced differentiation of the centrum while the neural tube induced differentiation of the neural arch. In 1955 he extended this study, using HH Stage 10–17 embryos and confirmed that removal of the neural tube resulted in an abnormally segmented cartilaginous sheath around the notochord, whereas removal of the notochord resulted in formation of an unsegmented cartilaginous gutter ventral to the spinal cord, associated with normally segmented neural arches. Notochord alone influences segmentation and in concert with the spinal cord influences differentation.

In 1954 Watterson et al. published a now classic paper on vertebral chondrogenesis in the chick. They addressed themselves to four questions: (a) what are the relative roles of neural tube and notochord in vertebral chondrogenesis; (b) if an induction occurs, how much association of inducer and competent tissue is required; (c) can any portion of the somite become sclerotome; (d) can various parts of the neural tube act as inducers? To answer these questions they used a variety of extirpation, implantation and grafting techniques. If they removed both the notochord and the neural tube from embryos with between 12 and 28 somites (HH 11–16), somitic chondrogenesis failed to occur, confirming that sclerotome cannot become chondrogenic in absence of notochord and spinal cord. If the notochord alone was removed, cartilage developed around the neural tube as a neural arch. If neural tube alone was removed, cartilage formed around the notochord as a centrum. Therefore both notochord and neural tube induce the cartilage which finally surrounds them, they function independently of one another and cannot spread their influences into the sclerotome adjacent to the other partner.

When somites were removed from 2-day-old embryos (HH 12–16) and grafted onto the CAM no cartilage formed (cf Murray and Selby, 1923). Somites from 3-day old embryos (HH 18–19) did chondrify on the CAM. Therefore association with notochord and spinal cord must have led to induction preceeding the third day.

Implantation of neural tube into the lateral face of the somite (potential dermomyotome) resulted in the induction of a neural arch, indicating lability of the somitic mesoderm.

Implantation of somites into the neurocoele was followed by chondrification. Therefore the ependymal lining of the neural tube has inductive ability similar to the outer face of the tube.

2.3. 1950's Onwards": In vitro Chondrogenesis

Grobstein and Parker (1954) and Grobstein and Holtzer (1955) were the first to attempt to culture somites in vitro. The fifth to the twelfth somites were isolated from foetal mice on the ninth day of gestation and cultured or grafted into the anterior chamber of the eye. No cartilage developed unless the spinal cord was included with the explants. Cartilage was the only tissue which developed in vitro. In those somites implanted into the eye, bone, haemopoietic tissue, fat and lymphoid tissue as well as cartilage differentiated — the environment influences to a considerable degree the type of differentiation which occurs. It was also shown (Grobstein and Holtzer, 1955; Flowers and Grobstein, 1967) that only the ventral half of the spinal cord possessed an ability to evoke cartilage from the somites, and then only for a limited time during embryonic development — spinal cord from twelve-day-old embryos was active, that from fifteen-day-old embryos was not. This polarity was reversed with respect to the ability of the spinal cord to evoke kidney tubule differentiation from intermediate mesoderm.

Ventral spinal cord from 4, $5\frac{1}{2}$ or $7\frac{1}{2}$ day-old chick embryos will induce chondrogenesis from somitic mesoderm taken from nine or twelve-day-old foetal mice (Grobstein, 1955; Cooper, 1965). The polarity was as found in the mouse and the ability to evoke chondrogenesis acted trans-species and trans-class.

These pioneering studies on in vitro chondrogenesis of mouse somitic mesoderm were quickly succeeded by studies using avian somites with or without spinal cord and/or notochord.

Lash et al. (1957) cultured somites from HH 17–18 chick embryos, in clusters of 6–8 somites on a liquid medium supplemented with horse serum and embryo extract and found that chondrogenesis was only initiated when spinal cord or notochord were included. They also showed that the spinal cord exerted an influence on both cartilage *and* muscle differentiation but that the notochord only influenced differentiation of cartilage.

Holtzer (1964b) obtained similar results — clusters of ten somites taken from HH 18 embryos and cultured for as long as three weeks failed to chondrify (although somites from HH 19 embryos produced cartilage in four days). Inclusion of ventral spinal cord enabled HH 18 somites to chondrify in four days. Dorsal spinal cord was ineffective — confirmed by Lash (1963a) but not by Strudel (1967).

The ability to chondrify and to respond to spinal cord or notochord was retained by the cells of dissociated somites (Stockdale et al., 1961). Somites from embryos of HH 13 to 19 were isolated, dissociated into individual cells and small clumps, pelleted and the pellets cut into isolates equivalent in size to the mass of 2 to 4 somites. If such isolates were associated with spinal cord or notochord chondrogenesis occurred; if they were left naked it did not. They also wrapped isolates of somitic mesoderm in ectoderm (see p. 25).

This phase of research established that pre HH stage 19 somites would not chondrify in vitro unless associated with spinal cord or with notochord.

These studies of the 1950's were followed by a series of studies, ongoing at the present time, which showed that isolated somites, free of spinal cord, notochord, neural crest derivatives, ectoderm or endoderm, *could* form cartilage (so-called *"spontaneous cartilage formation"*) if cultured in an appropriate medium — obviously a direct contradiction to the earlies studies of dependence on axial organs. The concept of environmental enhancement of a pre-established pattern within the somite — the expression of latent potential — began to appear in the literature.

Avery et al. (1956) appear to have been the first to recognise this phenomenon and to demonstrate it experimentally. They compared chondrogenesis in vitro with chondrogenesis expressed when somites were grafted to the chorio-allantoic membrane, with and without notochord or spinal cord; and the effect of donor age (Table 1). In

Table 1. Comparison of somites cultured on plasma clots with somites grafted to the choriollantoic membrane to show the H. H. stages at which spontaneous and "induced" cartilage are first observed (adapted from Avery et al., 1956). The % of clusters forming cartilage is also shown

Treatment	in vitro	on CAM
"spontaneous"	HH 18 (11 %)	HH 16 (17 %), HH 18 (86 %)[a]
with notochord	HH 16 (82 %)	HH 14 (76 %)
with spinal cord	HH 16 (87 %)	HH 14 (78 %)

[a]HH 18 shown for comparison with in vitro result

all situations cartilage formed from younger-aged embryos when the somites were grafted than when they were cultured, and more somites formed cartilage when grafted than when cultured. Thus whereas 86 % of HH stage 18 somites formed cartilage when grafted, only 11 % did so when cultured (Table 1). Notochord and spinal cord appeared to be equally effective in initiating chondrogenesis — either both are equivalent inducers, or both condition the somites or the somitic environment to the same extent. Doubt was shed on the concept that somites from pre-stage 18 embryos could only form cartilage when in contact with an inducer and on the concept that induction at that stage conferred some specificity to the somites.

Strudel (1962, 1963) showed that a variety of perturbations of the medium and of the in vitro environment could affect chondrogenesis from somites maintained free of spinal cord or of notochord. Thus he showed that somites from HH 16 embryos could chondrify if cultured in the presence of agar, embryo extract and horse serum. Somites from stage 12 embryos would also if first wrapped in pieces of vitelline membrane and then cultured with agar and embryo extract. Lash (1964) and Lash, Glick and Madden (1964) found that 10 % of cultures of HH 16–17 somites chondrified in presence of agar (vs 100 % in presence of spinal cord), and Zilliken (1967) obtained "spontaneous" chondrogenesis in 8 % of somites from HH 16 embryos. Without supplementation of the medium the somites dispersed and become fibroblastic (Cooper, 1965).

The mass of somites cultured is another viable which influences extent of chondrogenesis. On plasma clots, somites from embryos younger than HH 18 will not chondrify.

On agar somites from stage 16 embryos will chondrity. If 30 somites are cultured in a cluster then somites from embryos as young as HH 14 will chondrify (Holtzer, 1964a). Seventeen percent of cultures of somites from HH 13–14 embryos chondrify if cultured as single rows of somites, whereas 70 % chondrify if cultured with three or four rows of somites side by side. If the somites are first wrapped in vitelline membrane then the percentages are 37 and 84 % respectively (Thorp and Dorfman, 1967). Presumably we may attribute this enhancement to the protective and supportive conditioned environment created by contact with other tissues.

Ellison et al. (1969) have detailed information on the stage of development from which chondrogenesis can be initiated as a function of number of somites cultured (Table 2). The younger the stage the more somites required to allow chondrogenesis and the more somites cultured together for a given stage the greater the number of

Table 2. Effect of number of somites cultured and of age of donor embryo on percentage of cultures forming cartilage. Modified from data in Ellison et al. (1969)

HH stage	Number of somites cultured								
	1	2	4	6	8	12	16	24	32
	% Chondrogenesis								
9					0	0	0	0	0
10					0	0	0	0	11
11			0	0	3	37	40	33	14
12			0	0	19	12	33		
13			20	20	20	50	57		
14	0	0	33	56	69	90	100		
15	0	35	62	80	92	100			
16	27	43	79	91	100				
17	36	40	75	82	89				

Blanks are combinations of age and somite number not tested, Incidences of greater than 50 % of cultures chondrifying are circled

somites that chondrify. Similar results are obtained when somites are grafted to the chorioallantoic membrane (O'Hare, 1972a).

Only ten hours contact with the notochord or with the spinal cord is required to evoke chondrogenesis from somitic mesoderm (Holtzer, 1964a), whereas continuous contact with the enriched medium is necessary to initiate spontaneous chondrogenesis (Lash, 1968a). Lash further postulated that one hundred percent of the somites cultured from embryos of any age would chondrify if the appropriate conditioned medium were found and used, although he emphasised (1967) that the potential within somites from various levels of the embryonic axis was not equivalent with respect to ease of expression of this "spontaneous" chondrogenesis. Kosher (1976) has provided evidence that the somite itself might undergo self-induction by positive feedback from products of the extracellular matrix and that this is the basis of spontaneous chondrogenesis. This is quite a shift in position from the early studies of complete dependence of somite mesoderm on information received from an inducer. The timing is important in these studies. Somites posterior to numbers 22–24 in HH stage 15 to 18 embryos form more cartilage (70 % of cultures) than do the anterior somites (15–20 % of cultures) when cultured alone but all cultures chondrify if spinal cord is added (Lash, 1967; Gordon and Lash, 1974).

The early workers (Williams, 1910), recognized that the somites arose and chondrified in an anterior-posterior sequence which reflected the anterior-posterior developmental gradient of the embryo as a whole. It was therefore surprising when Lash (1967) observed that 80 % of posterior somites, but only 20 % of anterior somites, from HH stage 16 embryos chondrified in vitro, in the absence of the notochord. The anterior somites from stage 17 embryos only formed cartilage if the notochord was present, perhaps because they had too little contact with the notochord in ovo or because their histo-architecture did not readily permit survival in vitro, (Gordon and Lash, 1974).

The normal in vivo anterior-posterior gradient is expressed when somites are grafted onto the chorio-allantoic membrane (O'Hare, 1972a). The further anterior in the em-

bryos and the older the embryo the better the incidence of cartilage on the CAM
For every stage studied the clusters of four posterior somites produced fewer nodules of cartliage than did the four somites immediately anterior to them. This would seem to eliminate the possibility of insufficient contact of anterior somites with the notochord, and perhaps indicate that the morphology of the anterior somites is insufficient for the transfer to in vitro conditions.

One of the most detailed studies on effect of environment on chondrogenesis is that by Ellison et al., (1969). They cultured strips of eight of the 10 most posterior somites from embryos of HH stages 9 to 17 in a variety of supplemented media. In the presence of horse serum HH 16 somites chondrified (13 % of cultures). With foetal calf serum three percent of HH stage 11 somites chondrified, and if the conditioned medium was concentrated by overlaying with a layer of paraffin, 37.5 % of HH 9 somites chondrified. Thus as early as HH stage 9, somitic mesoderm can express chondrogenic potential. Subsequently they showed (Ellison and Lash, 1971) that somites from HH 17 embryos cultured in the presence of foetal calf serum; produced as much cartilage as early as did somites cultured in the presence of the "inducer". The spinal cord and notochord are not unique in their action, or in the degree of action, on somitic mesoderm.

Increasing the K^+ concentration of the medium from 2.7 mM to 4.9 mM increases the amount of glycosaminoglycan synthesized by the somites during the first twenty four hours in culture but decreases the number of cultures which form cartilage by 20 % (Lash et al., 1973). However addition of glycosaminoglycan to the medium increases the amount of cartilage obtained (Kosher, et al., 1973). It is evident that we have a lot to learn about factors which potentiate chondrogenesis, and of the conditioning effect which they have on the medium (see pp. 36–37).

The agent BUdr (5-bromo-2'-deoxyuridine) blocks the terminal differentiation of a wide variety of cell types. It as an analogue of thymidine, substituting for thymidine in the treated cells. Abbott et al., (1972) exposed 10 of the posterior somites from HH 17–18 embryos to BUdr (10 µg/ml medium) for three-day periods and then cultured the dissociated cells with or without the notochord for a further 10–25 days. Their results are reported in the table below.

Exposure to BUdr during the first three days of culture prevented the appearance of cartilage nodules after further incubation in the absence of BUdr but in the presence of excess thymidine. The mechanism responsible for this inhibition remains undetermined – a block to mitosis, selective cell death, a block to chondrogenic determination and inability of cells to synthesis or export chondroitin sulphate have all been postulated (Holtzer et al., 1972; Holtzer and Mayne, 1973).

Table 3. Number of cultures with cartilage 10–25 days after exposure to BUdr

Days in BUdr	somites + notochord		somites alone	
0–3	0/48	0 %	0/12	0 %
1–4	0/12	0 %	–	
2–5	3/ 9	33 %	–	
3–6	5/ 5	100 %	–	
control	44/44	100 %	29/32	91 %

Holtzer regards cell division as of primary importance in the expression of chondrogenesis from somitic mesenchyme. He stresses that the initial induction around HH stage 12 occurs some 30–40 hours before cartilage is first seen, and therefore it is the progeny of induced cells which actually chondrify. The rate of cell division and the number of cells produced is higher in somites cultured with notochord than in those cultured alone (Holtzer, 1964b; Holtzer and Mayne, 1973). The notochord, and also agents within permissive media, such as foetal calf serum, may act by minimizing cell death as well as by enhancing cell division (although Gordon and Lash, 1974 have argued that the notochord does not stimulate the rate of mitosis within somitic cell but rather than it minimizes loss of DNA). They have preliminary evidence from in vitro studies that the pre-cartliage cells are more susceptible to cell death than are other somitic cells, that the notochord minimizes such death, thereby allowing accumulation of a critical mass of determined chondrogenic cells for chondrogenesis to commence. However cell death is only infrequently seen within the somite in vivo and then only at some distance from the notochord (Minor, 1973).

3. The Inductive Tissues

Although we have just summarised the literature which indicates that neither the notochord nor the spinal cord provide information to the somites which is essential for somitic chondrogenesis to commence in vitro, it is evident that these tissues do provide developmental information to the somites in vivo. In vitro we are able to by-pass the inducer by using conditioned medium but the embryo is not able to compensate for the absence of the spinal cord or of the notochord, so that in vivo these tissues are prerequisites for somitic chondrogenesis. There seems therefore, to be justification for continuing to refer to them as inducers and to their interaction with the somites as an inductive interaction.

In this section we will explore what is known of the inducers and of their mechanism of action on the somites.

3.1. Integrity of the Inducer and Vertebral Morphogenesis

Must the inducer be intact and morphologically normal for it to induce cartilage from somitic mesoderm? Watterson (1952) showed that normal neural arches developed even in the presence of disorganized neural tissue in *Fundulus heteroclitus*. In the urodele amphibia, on the other hand, abnormalities of the spinal cord lead to abnormalities of the vertebral column (e. g. Detwiler and Holtzer, 1956) as they do in avian embryos. Administration of nicotine sulphide to 2-day-old embryos results in the notochord of the neck becoming twisted within hours and in vertebral deformation several days later (Strudel and Gateau, 1971).

If the inducing organ is *mechanically* injured it can still induce cartilage which is normal by biochemical or ultrastructural criteria, but such cartilage does not organise into a morphologically normal spinal column. *Chemical* disruption of the inducer renders it unable to induce cartilage at all (Strudel, pers comm.).

There is also a difference in the integrity of the *cartilage* induced which depends on whether the spinal cord or the notochord was used as the inducer. Lash (1968b) found that on either liquid or agar media, spinal cord-induced cartilage dissipated between the ninth and sixteenth days in vitro whereas notochord-induced cartilage persisted intact until at least the twenty first day in culture. He suggested that the spinal cord plays a role in remodelling the vertebral column as it grows in vivo, thereby preventing compression of the developing spinal cord. This interesting idea has not been pursued further.

O'Hare (1972c) has conducted an interesting experiment to test whether viability of the spinal cord is a requisite for induction. If he grafted the last four somites of HH 9–12 embryos onto the chorioallantoic membrane in the presence of intact spinal cord, cartilage was produced (95 % of the grafts). If the donor embryos were irradiated with 5000 R before removing the spinal cord for grafting, cartilage still formed in 52 % of the somites. He suggested that "two day spinal cord . . . . possesses . . . . specific chondrogenic activity that survives a lethal dose of X-radiation". But the induction may have occurred in the first hours of contact of irradiated spinal cord and somite before cell death of the inducer was complete.

3.2. When Do the Inducers Lose Their Inductive Ability?

As the spinal cord and notochord appear to have a specific role to play in induction of somitic cartilage it is of interest to know whether they ever lose their inductive ability. We may summarise by saying that the notochord, as late as it has been tested, retains inductive properties but that the spinal cord loses them.

Cooper (1965) showed that the notochord from embryos as old as HH 37.5 (11 $^1/_2$ days) could induce somites from HH 17 embryos or from 9-day-old foetal mice to form cartilage in vitro. He did not test older ages of notochord. Notochord from nine-day-old mice would also induce.

The spinal cord does not retain inductive ability for the same period. This was tested, again by Cooper, by placing spinal cord from chick and mouse embryos in contact with somites from nine-day-old mouse embryos. Spinal cord from 7 $^1/_2$ day-old chick embryos would induce, that from 9-day-old would not, although we have recently found that spinal cord from embryos as old as H. H. stage 44 will allow H. H. stage 17 somites to chondrify in vitro under conditions where somites alone fail to chondrify (Tremaine, R, unpublished results). Spinal cord from nine-day-old mice was effective Grobstein and Holtzer (1955) showed that spinal cord from nine and twelve day, but not from fifteen day-old mice could induce. The data to date are summarised in Table 4.

O'Hare (1972b) used spinal cords from nine-day-old chick embryos to test whether HH stage 9–12 somites grafted to the chorioallantoic membrane would form cartilage. The absence of cartilage was attributed to the inability of the somites to respond to the inducer. It is evident that the spinal cords used were no longer inductively active.

Table 4. Latest known age for inductive ability

	spinal cord	notochord
chick	between $7\frac{1}{2}$ – 9 days	after $11\frac{1}{2}$ days[a]
mouse	beween 12–15 days	between 12–15 days

[a] older foetuses have not been tested

Inductive ability and its loss may also be correlated with the state of differentiation of the inducer. Cooper's (1965) general conclusion was that notochord (and cartilage, p. 28) only induce when their cells are undergoing hypertrophy or vacuolation [see Bancroft and Bellairs (1976) for ultrastructure]. This conclusion was based on organ culture (either direct or trans-filter) and the results are summarised in Table 5. Absence of hypertrophy correlated with loss of inductive ability and onset of vacuolation correlated with initiation of inductive ability.

Table 5. Incidence of cartilage in somites maintained in vitro as a function of state of cellular differentiation of notochord. Based on data in Cooper (1965)

Notochord			Somite			
source	age	state[a]	source	age	Percent of cultures with cartilage	
mouse	9 day	1	mouse	9 day	15/18	(83 %)
chick	HH 12–15	1	chick	HH 15–18	10/12	(83 %)
chick	HH 18	2	chick	HH 18	20/20	(100 %)
chick	HH 21–24	3	mouse	9 day	18/18	(100 %)
chick	HH 27–$37\frac{1}{2}$	4	mouse	9 day		
			chick	HH 17	77/77	(100 %)
chick	HH $37\frac{1}{2}$	5	mouse	9 day	0/ 2	(0 %)
			mouse	9 day	0/29	(0 %)
			chick	HH 14, 15	0/26	(0 %)
			chick	HH 17, 18	0/26	(0 %)

[a] state of differentiation of notochord:
1 non-vacuolated, 2 initial vacuolation, 3 later vacuolation, 4 hypertrophic, 5 non-hypertrophic

3.3. Ability of Dermomyotome to Chondrify

During normal development the only portion of the somite to chondrify is the sclerotome. The question has been asked: does this reflect a chondrogenic bias of the sclerotomal mesenchyme not shared with the remainder of the somite or could dermotome and myotome form cartilage if suitably challenged? The latter view would hold that their distance from the normal inducers is all that normally prevents them from chondrifying. Experiments have been carried outh both with amphibia and with birds.

Holtzer and Detwiler (1953) removed 75 % of the medial aspect of the somites of tail bud-stage urodele embryos and found that cartilage differentiation proceeded normally although muscle development was delayed. Apparently areas of lateral somite (myotome, dermotome) become chondrogenic rather than myogenic so that chondrogenesis took precedence over myogenesis. Implantation of spinal cord or of notochord into the lateral face of the somite (i. e. into presumptive dermo-myotome) led to the formation of a secondary neural arch in this ectopic site, indicating that both differentiation and morphogenesis of the cartilage into a segmented structure could occur (Holtzer and Detwiler, 1953 for amphibia; Strudel, 1953b for chick).

Cooper (1965) placed myotome from HH stage 18 chicks with notochord and spinal cord (age unspecified) in vitro without observing chondrogenesis. However, it may well be that by stage 18 the myotome is fixed with respect to ability to become chondrogenic.

O'Hare (1972c) grafted, to the CAM of 9-day old embryos, lateral mesoderm from a site adjacent to the last four pairs of somites of HH stage 9–12 embryos. His results are presented in Table 6 and indicate that lateral mesoderm from the chick certainly has the ability to respond both to the spinal cord and to the adjacent epithelia by forming cartilage. This experiment seems to remove the interpretation applied to the earlier experiments that the transplanted inducer acted, not to induce lateral mesoderm, but to attract sclerotome cells to the lateral mesoderm where they formed the cartilage.

Table 6. Formation of cartilage in lateral mesoderm when placed in contact with spinal cord or with epithelia, and grafted to the chorioallantoic membrane of a host embryo. Data taken from O'Hare (1972c)

Treatment	Grafts with Cartilage (%)
1. Lateral mesoderm alone	4 %
2. Lateral mesoderm + ectoderm or endoderm	37 %
3. Lateral mesoderm + spinal cord	42 %
4. Lateral mesoderm + irradiated spinal cord	2 %

3.4. The Search for a Pure Inducer

The search for the pure inducer which has pervaded many other areas of developmental biology and embryology has not been passed over by students of somite chondrogenesis. For a brief ten-year period (1953–1963) it looked as if progress was being made. Nothing has happened since then.

Strudel in a series of papers (1953b, 1959, 1962) established that a saline extract of the spinal cord or notochord of 3–3½ day-old embryos would, if placed in vitro with stage 13 or older somites, induce them to chondrify. In 1963 he extended this back in time to stage 11 somites. There it rests ... Strudel concluded: "even very young somite cells are genetically determined cells and all that they need to undergo or accomplish their phenotypic differentiation is a microenvironment favoring chondrogenesis. This does not mean that the inducing action of the spinal cord and the notochord is dispensable. It may be that the spinal cord and/or the notochord exercize their inducing effect very early ... " (Strudel, pers comm.).

Lash et al. (1957) set up the following experimental procedure. Spinal cords were placed trans-filter to somites in vitro for 10 hours. The spinal cord was then removed and the somites left to develop in isolation. Three days later cartilage had differentiated within the somites. Cartilage was not found in somites cultured in isolation from the start or in those cultured adjacent to the "spinal-cord spot". His conclusions were (1) only a short period of contact between somite and inducing spinal cord is needed to effect the evocation of cartilage several days later; (2) many undetected changes must have occurred within the somite between the removal of the spinal cord and the appearance of the cartilage; (3) transfer of a chemical had occurred between spinal cord and somite. In subsequent experiments (Lash, 1968a) fragments of spinal cord were placed on agar for 24 hours, the spinal cord removed and replaced with clusters of somites. More cartilage was found in these somites than in those placed elsewhere on the agar.

In 1962, Lash et al. and Hommes et al. extracted a perchloric acid extract from the spinal cord and notochord of $2 \frac{1}{2} - 3$ day old chick embryos which allowed somites to chondrify in vitro. They found that somites from stage 16 embryos failed to form cartilage in vitro if in isolation, but that 90 % of them chondrified in the presence of the extract. Only 14 % of "unstimulated" somites from HH stage 17–18 embryos form cartilage in vitro but 82 % did so when cultured with the extract. This fraction enables initiation of chondrogenesis in "dormant" somites and enhances chondrogenesis in activated somites. The extract contained nucleotides, amino acids and carbohydrate and it was evidently the nucleotide fraction which was the active one. Zilliken (1967) isolated an oligopeptide fraction from 1000 trunks (notochord, spinal cord, somite and epidermis) of $4 \frac{1}{2}$ day chick embryos, which allowed 50 % of HH stage 16 somites to chondrify in vitro. This extract both increased the incidence of cartilage nodules and increased the uptake of S^{35} by the treated somites.

The early work on the search for informational molecules as mediators of chondrogenesis has been reviewed by Lash (1963a). It is worth noting that Dorfman (quoted in Thorp and Dorfman, 1967), Holtzer (1961), and Holtzer and Mayne (1973) were unable to repeat these experiments and that no other labs have obtained independent verification of the isolation of an active inductive fraction from notochord or from spinal cord.

4. Ectopic Inducers

4.1. Can Ectoderm Act Inductively on Somites?

As mentioned in section 2.1. (pp. 13–14) the early studies in the 1920's, 30's, and 40's which involved grafting of somites dissected from chick embryos onto the CAM involved the transfer or adjacent ectoderm and possibly of subjacent endoderm within the grafts. This problem has been taken up by later workers in an endeavour to determine whether the ectoderm which normally overlies the developing somite plays any role

Table 7. Summary of the results of experiments testing the ability of heterologous potential inducers for ability to evoke chondrogenesis within somitic mesoderm[a]

Tissue tested for inductive ability	Test Situation	Reference
MOUSE *Active*		
9 or 12 d fore-brain, 9 d medulla, 12 d mid brain, 11 day otic capsule epithelium	six, 9-day somites in vitro	
Inactive		Grobstein and Holtzer (1955)
15 d rib cartilage, 15 d liver, adult liver, kidney, spleen	six, 9-day somites in vitro	
CHICK *Inactive*		
1. 3–3½ d brain, gut, mullerian duct	(a) 4–12 somites of HH 10–17 in vitro (b) transplant to replace notochord and spinal cord in ovo.	Strudel (1955, 1962)
2. 3 d liver, sensory ganglia, epidermis, muscle	ten, HH 18 somites in vitro	Holtzer (1964b)
3. 18 d liver, kidney, muscle	six, HH 16 somites in vitro	Avery et al. (1956)
4. 3 d otic vesicle epithelium	(a) HH 15–23 somites in vitro (b) transplant to replace HH 12–13 notochord in ovo.	Benoit (1960)
5. 3 d optic or otic vesicle, 4 d myocardium, mesonephros, 9 d ventricle, liver, intestine, choroid plexus, forebrain, Polyoma-transformed fibroblasts	four, HH 9–12 somites on CAM	O'Hare (1972b)
6. epidermis	HH 18 somites in vitro	Holtzer, (1964b)
7. ectoderm	HH 18 dissociated somites in vitro	Stockdale et al. (1961)
8. ectoderm + endoderm	HH 14–18, in vitro or grafted intracoelomically	Lash (1963b)
9. 3 d trunk epidermis	HH 9–12 somites in vitro	O'Hare (1972b)

Table 7 (continued)

Tissue tested for inductive ability	Test Situation	Reference
Active		
1. ectoderm + endoderm	HH 10–13 somites on CAM	Seno and Buyuközer (1958)
2. ectoderm + endoderm 4 d limb ectoderm 3, 4 d trunk ectoderm	HH 9–12 somites on CAM	O'Hare (1972b)

[a]Cartilage as a heterologous inducer is shown in Table 8

in the induction of the cartilage which forms in the sclerotome. The results are contradictory and the final story is not yet told. However, it is evident that no action is exerted by these epithelia in vivo for chondrogenesis fails to occur after extirpation of the spinal cord and/or notochord. The epithelia cannot overcome this loss.

Holtzer (1964b) notes that among the tissues tried for inductive ability in his study was epidermis (source and age not specified) and that it would not induce HH stage 18 somites to chondrify in vitro (Table 7).

Seno and Büyuközer (1958) grafted somites from 9–20 somite chicks (HH 10–13) to the CAM with and without the overlying ectoderm and underlying endoderm. In the absence of either epithelium no cartilage was found (0/16). In the presence of both ectoderm and endoderm 75 % (30/40) of the grafts produced cartilage. In presence of ectoderm alone, 57 % (8/14) produced cartilage whereas in presence of endoderm alone 14 % (2/14) produced cartilage. They concluded that the epithelia, especially the ectoderm provide support for the sclerotomal mesenchyme, preventing cell dissipation, and that some physiological intervention was probably also provided. Stockdale et al. (1961) tested the ability of the ectoderm to induce chondrogenesis from pellets of somitic cells – an odd protocol if cell stability is important. They isolated somites from stage 18 embryos, dissociated the somites into individual cells or cell clumps, pelleted the cells by centrifugation and then cut the clumps into pieces equivalent in size to 2–4 somites. If these were clustered around notochord or spinal cord and cultured, cartilage differentiated. If cultured around ectoderm, no cartilage was found. This negative result should not lead us into thinking that the experiment rules out a role of ectoderm in maintaining tissue relationships in *intact* somites in vivo.

Lash (1963b), cultured or grafted intracoelomically, either strips of somites with ectoderm or endoderm intact, or clusters of somites to which ecto-or endoderm has been added. These were taken from HH stage 14–18 chick embryos and in 90 cultures no cartilage was found (confirmed by unpublished results of Strudel, pers comm.).

Then O'Hare (1972b) took another look at the question of ectodermal involvement in somite chondrogenesis, using an in ovo CAM-grafting technique (Table 7). The last four somites from stage 9–12 embryos were grafted to the CAM on millipore filters in association with a variety of types of ectoderm. Somites grafted alone failed to chondrify. Somites grafted with adjacent ectoderm *and* endoderm formed cartilage in 21 % of the grafts. Somites reassociated with ectoderm and endoderm after trypsinization to separate the components, formed cartilage in 10 % of the grafts. Thus there was a definite association of the epithelia normally adjacent to the somites and evocation of chondrogenesis (albeit in a low percentage of cases) from somites. O'Hare also showed that the ability to augment was not restricted to epithelia adjacent to somitic mesoderm but was present in ectoderm from limb buds (this ectoderm included the apical enctodermal ridge) of 4-day embryos (24 % of grafts) and also from trunk ectoderm of specific ages. Thus trunk ectoderm from 2-day embryos failed to act; that from 3-day embryos allowed cartilage production in 30 % of the grafts, and that from 4-day embryos allowed cartilage in 23 % of the grafts. O'Hare related this ability of the epithelia to evoke chondrogenesis to the presence of a basement membrane and to the extracellular matrix in these epithelia (see p. 37). Lateral plate mesoderm could also be induced to chondrify (37 % of grafts) if grafted with ectoderm from adjacent areas (Table 6).

No further studies on the role of these epithelia have been carried out. The tentative conclusion which I draw from the existing experiments is that ectoderm will allow somites to chondrify in the absence of notochord or spinal cord, if such somites are grafted to the chorioallantoic membrane but not if they are maintained in vitro. The only conclusion which we can draw concerning any role of the ectoderm in vivo is that it cannot act in the absence of spinal cord or of notochord (the extirpation experiments).

4.2. Other Ectopic (heterologous) Inducers

Table 7 summarises the results from experiments using other heterologous potential inducers. A considerable variety of tissues have been tested, from both mouse and chick embryos of various ages, and in various experimental situations — in vitro in combination with somites, as grafts with somites to the CAM, or as in ovo transplants to replace the excised notochord or neural tube. The results obtained do, in some measure, depend on the species involved. Thus neural ectodermal derivatives and otic capsule epithelium from foetal mice will induce mouse somitic mesoderm to chondrify in vitro, (Grobstein and Holtzer, 1955) but the equivalent tissues from the embryonic chick fail to elicit chondrogenesis, either in vitro or in ovo (Strudel, 1955, 1962; Benoit, 1960; O'Hare, 1972b — see Table 7). Why the mouse should possess this "specific inductive ability" is unknown. In normal life they presumably induce the overlying mesenchyme to ossify as does the neural tissue of the chick (Benoit and Schowing, 1970, for review). Inability of the otic epithelium of the chick to induce somitic mesoderm to chondrify is not due to inactivation of inductive ability because of the ectopic site for otic epithelium will induce otic mesenchyme to chondrify in the same ectopic sites (Benoit, 1960). The specificity may be in the responding mesenchyme. Such tissues which induce cartilage to form from one mesenchyme but not from another warrant much closer study as do the responding mesenchymes.

The unequivocal results obtained from all the tissues of the chick except epithelia and cartilage (to be discussed below) is that they fail to elicit chondrogenesis from isolated somites in vitro even though the somites and the test-tissue remain viable and continue to grow (Table 7). It therefore appears that the tissues adjacent to the sclerotome, viz. the notochord and the spinal cord as well as epithelial tissues, do have particular abilities not possessed by other tissues with respect to enhancement of somitic chondrogenesis. Having said that, there is one exception to this rule. That is that chondroblasts and chondrocytes from a variety of skeletal sites (ribs, long bones,

Table 8. Incidence of cartilage in somites cultured trans-filter to cartilage and dependence of induction on stage of histogenesis of cartilage. (Data from Cooper, 1965)

Stage of cartilage differentiation[a]	Incidence of cartilage	
Mouse		
1	0 %	(0/51)
2–3	91 %	(51/56)
3–4	0 %	(0/12)
Chick		
1	0 %	(0/17)
2	93 %	(27/28 mouse somites)
		(15/17 chick somites)
3	21 %	(3/14)
4	95 %	(20/21)

[a]Stage 1, small celled, round chondroblasts; stage 2, flattened chondroblasts; stage 3, enlarging chondrocytes; stage 4, terminal hypertrophy (mouse), loss of hypertrophy (chick)

vertebrae, trachea), can, at particular cytological stages in their differentiation, induce somitic mesoderm to chondrify. Cooper (1965) has established this relationship using trans-filter cultures of mouse and chick somites and cartilage. Cartilage from 12 to 15 day old mouse foetuses or from 5 $\frac{1}{2}$ to 15 $\frac{1}{2}$ day-old embryonic chicks was placed transfilter to either 8 somites of 9-day mice, 8 somites of HH 12–15 chicks, or to 6 somites of HH 17–18 chicks. The results are shown in Table 8. Only chondrocytes which are enlarging or becoming hypertrophic acted as inducers. Thus with stage 2–3 cartilage from the mouse (see Table 8), induced somitic cartilage was found adjacent to flattened or enlarging cells but not adjacent to fully hypertrophic chondrocytes (the latter had completed their hypertrophy before being cultured). Stage 1 chondroblasts fail to become flattened chondroblasts in vitro and fail to act as inducers. Cartilage of stages 3–4 from the mouse remains "terminally hypertrophic" in vitro and fails to induce whereas stage 4 cartilage from the chick undergoes loss of hypertrophy and cytolysis in vitro and does induce. This association of inductive ability with onset of chondrocyte hypertrophy and loss of ability with attainment of hypertrophy is analogous to the attainment and loss of inductive ability by the notochord (p. 22). It is curious that the tissue to be induced (cartilage) is the tissue which has the ability to induce somites to chondrify. It may be even more instructive that the particular extracellular matrix products of the induced cartilage are strikingly similar to those produced by the inducing cartilage and by the normal inducers, spinal cord and notochord (pp. 31–36).

5. Synthesis of Glycosaminoglycans and Onset of Cartilage Differentiation

The production of glycosaminoglycans, especially chondroitin sulphate, is characteristic of functioning chondroblasts and chondrocytes. The onset of such synthesis, as detected by ^{35}S autoradiography, has been used as a means of detecting the initial appearance of cells determined for chondrogenesis.

Amprino (1955) showed that there was uptake of S^{35} into prechondrogenic somites of both chick and mammalian embryos. Johnson and Comar (1957) confirmed this finding and emphasised that maximal uptake occurred in the mesenchyme immediately adjacent to the axis. Lash et al. (1960), apparently unaware of this study, sought to determine whether glycosaminoglycan accumulation preceeded histological differentiation of cartilage and found that they could not detect chondroitin sulphate earlier than they could detect metachromatic extracellular matrix. They did find incorporation of S^{35} preceeding cartilage differentiation but felt that this was a precursor of chondroitin sulphate [subsequently shown to be sensitive to RNA'ase, Lash (1963a)].

Somites explanted in vitro without the spinal cord or notochord fail to chondrify and also fail to produce collagen or chondroitin sulphate. Glick et al. (1964) cultured somites from HH stage 16–17 embryos for 8 days without observing any activity of the enzymes ATP-sulphurylase or APS-kinase (both of which are involved in the sulphation of Chondroitin), unless the notochord or spinal cord were cultured along with the somites — enzyme induction in vitro?

With refinements in techniques it became possible to detect chondroitin sulphate before histochemically-detectable extracellular matrix could be seen. For example, Franco Browder et al. (1963) in their study of somites, notochord and spinal cord from HH stage 11–23 embryos observed the production of chondroitin sulphates A and C even from the stage 11 tissue. It was becoming apparent that, based on synthesis of chondroitin sulphate, cells of the somitic mesoderm are determined for chondrogenesis long before extracellular matrix is detectable (Okayama, et al., 1976). Further developments in methodology produced a thin-layer chromatographic technique capable of isolating the precursors of chondroitin sulphate (Marzullo and Lash, 1967a, 1967b). It then became apparent that many tissues of the young embryo (somite, epidermis, spinal cord, notochord, mesonephros, extra-embryonic membranes) could metabolise glucosamine and form UDP-N-acetylgalactosamine (UDP-NaGal) but that in these tissues the amount of UDP-NaGal produced was very small in comparison with that produced by cartilage. Thus the concept arose of a somite fully able to produce chondroitin sulphate, but needing a stimulus to augment and stabilise the rate of synthesis and to enable it to accumulate extracellular maxtrix (Lash, 1968b). Many tissues can synthesize chondroitin sulphate but not all can accumulate it (Abrahamsohn et al., 1975).

Toole (1972) has studied the relationship between hyaluronate and chonroitin sulphate within the chondrifying somite. He studied somites 20–32 (those lying between the wing and the leg) in embryos of HH stages 23–30. The ratio of hyaluronate to chondroitin sulphate was very high at stage 23 and very low thereafter (Table 9), the decrease coinciding with increased levels of hyaluronidase (Table 9) and with the formation of metachromatic matrix in the peri-notochordal mesenchyme (stage 25). Removal of hyaluronate apparently accompanies formation of extracellular matrix.

Table 9. Ratio of hyaluronate: chondroitin sulphate and levels of activity of hyaluronidase within somites of chick embryos between HH stages 23 and 30. (Modified from Toole, 1972)

H. H. Stage	Hyaluronate: chondroitin sulphate	Hyaluronidase[a]
23	1.9	12.5
24		48.1
25		45.0
26, 27	0.6	34.0
28	0.4	44.3
30	0.2	31.2

[a] nanograms of N-Acetylhexosamine released/g protein/hour by 100 μg, hyaluronate at 37°

Exposure of somites to spinal cord or to notochord does not confer upon them the ability to synthesis chondroitin sulphate and hyaluronate but enhances the rate at which pre-existing synthetic pathways operate. The instruction from the "inducer" is to speed-up not to start-up.

6. Production of Extracellular Matrix by Notochord and by Spinal Cord

The evident role which notochord and spinal cord play in evoking somitic condrogenesis in vivo, and the accumulating evidence that normal constituents of the extracellular matrix of cartilage (chondroitin sulphate, collagen, hyaluronate) could influence the rate of synthesis of cartilage extracellular matrix by feedback inhibition and stimulation (Hall, 1973), lead to a search for the existence of extracellular matrices around the notochord and spinal cord. The first studies bearing on this problem utilized histochemical and autoradiographic techniques to show that glycosaminoglycans could be visualised around the notochord at the time when chondrogenesis was commencing within the adjacent sclerotome.

6.1. Glycosaminoglycans

Johnson and Comar (1957) injected S^{35} into the albumen of the egg and detected uptake into the notochord and the primitive streak at stages 3+ and beyond, i. e. long before somite chondrogenesis. Franco-Browder et al. (1963) confirmed that notochord and spinal cord from HH stage 11 embryos was surrounded by sulphated glycosaminoglycans. Lash (1963) and Lash et al. (1964) noted S^{35} uptake over the spinal cord, notochord and immediately adjacent extra cellular matrix in vitro but were unable to judge where the material was produced. O'Connell and Low (1970) using electron miscroscopy and Kvist and Finnegan (1970a) using histochemistry, detected sulphated glycosaminoglycans adjacent to the notochord at stages 16 and 17 respectively. Kvist and Finnegan (1970b) went on to study the hyaluronic acid: chondroitin sulphate ratios within the axis. At stages 17, there was 2 $\frac{1}{2}$ times more hyaluronic acid than chondroitin sulphate, whereas by stage 28 this ratio had been reserved. They suggested that the hyaluronate played a role in sclerotome aggregation (see p. 9). Frederickson and Low (1971) noted amorphous material (GAG?) on microfibrils around the notochord of 2-4 day embryonic chicks (Fig. 4). Strudel (1971) and Corsin (1974) emphasized the importance of the metachromatic extracellular matrix which they found between the sclerotome and the nervous system and around the notochord and though that it progressively became the matrix of the perinotochordal cartilage. Minor (1973) and Bancroft and Bellairs (1976) using electron microscopy, observed the accumulation of 200-400 Å glycosaminoglycan granules along with 150 Å unbanded collagen fibrils around the notochord and ventral part of the spinal cord at HH stage 10 (33-38 hours). It is relevant in the light of the inductive role played by the ventral spinal cord but not by the dorsal cord (p11) that matrix products were only found around the *ventral* portion of the spinal cord. By HH 17 this extracellular matrix was very prominent with a basement membrane surrounding the notochord and spinal cord. This maturation of the matrix preceeds the migration of the sclerotomal cells around the notochord, which possess occurs at HH 18.

The extracellular matrix and the basement membrane are lost when trypsinization is used to isolate somites. Evidence that matrix products will reform when notochord or spinal cord are maintained in vitro provides direct evidence for the production of the matrix by these tissues. When segments of the trunks of 3-day old chick embryos, consisting of 9-10 somites, notochord and spinal cord, are maintained in vitro in Ham's F-12 + 10 % FCS, for 24 hours in the presence of $^{35}S_4$, labelled products are found in the extracellular matrix. Enzymatic digestion of the products from the cultures of *isolated* notochords and neural tubes indicates that the bulk of the products of both notochord and neural tube are chondroitin sulphate and heparin sulphate (Hay and Meier, 1974). The comparative biochemistry which has been done indicates homology of structure between notochord and cartilage chondroitin sulphate-protein complexes (Mathews, 1971, 1975).

Thus both the notochord and the ventral portion of the spinal cord synthesis and accumulate glycosaminoglycans into an extracellular matrix at least as early as HH stage 10 in the embryonic chick.

6.2. Collagen

From early 1960's onwards it became accepted that epithelial structures such as notochord and spinal cord might have the ability both to synthesise and export collagen.

Jurand (1962, 1974) and Bancroft and Bellairs (1976) described the ultrastructure of the notochord in the embryonic chick and mouse and described fibrils in a perinotochordal position. They thought that some were derived from adjacent sclerotome cells and that some came from the notochordal sheath. Then came a series of studies by Low and his colleagues. O'Connell and Low (1970) and Frederickson and Low (1971) described microfilaments both on the notochord, and between the notochord and the spinal cord, of the embryonic chick as early as HH stage 11. In specimens from

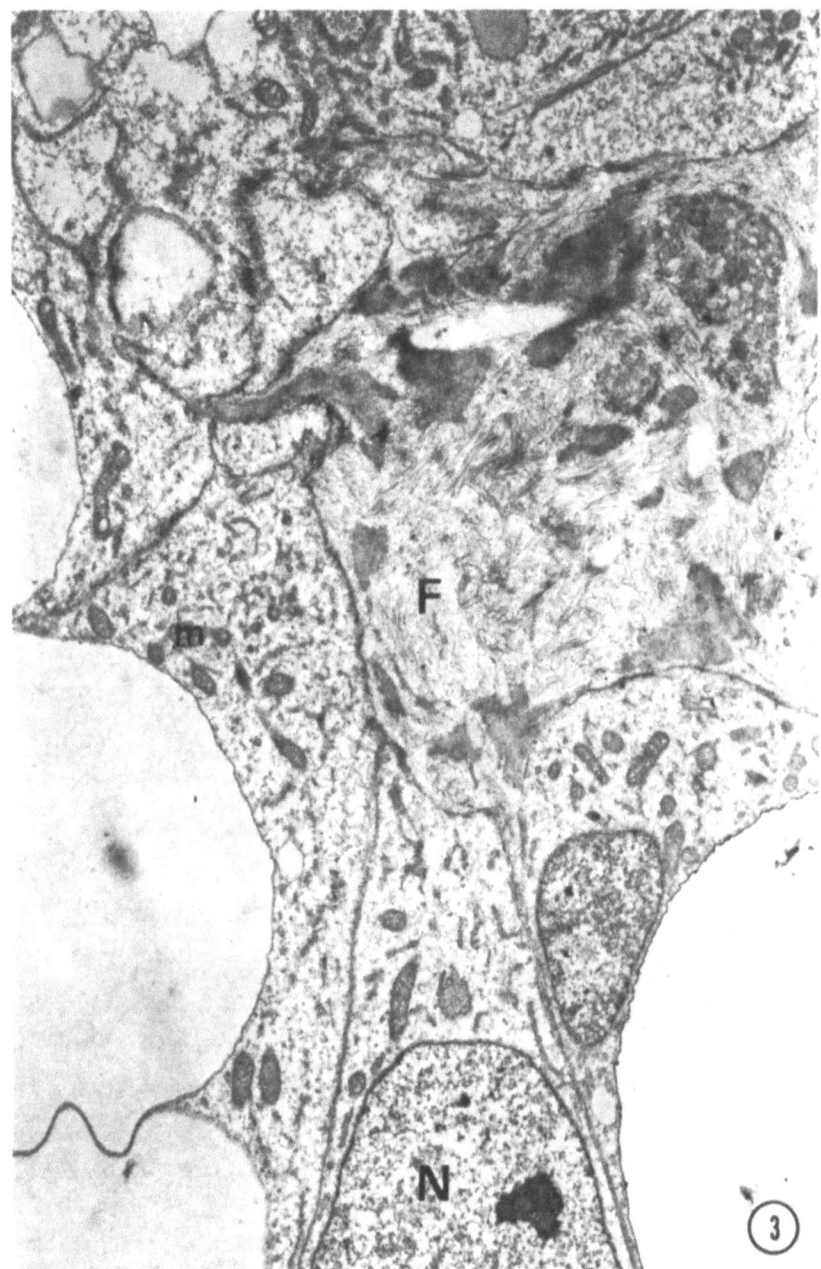

Fig. 3. A low power (X 7,500) electron micrograph of developing notochord of embryonic chick. Note the dense accumulation of extracellular fibrous material (*F*). *N*, nucleus; *m*, mitochondrion. (Courtesy of Dr. G. Strudel)

2–4 day-old embryos these microfibrils were observed to have amorphous material (interpreted as glycosaminoglycans) attached to them. The microfibrils were of two types: (a) 150–200 Å diameter, unbanded, beaded fibrils close to the notochord and sensitive to removal by collagenase, (Minor (1973) has described similar fibrils from

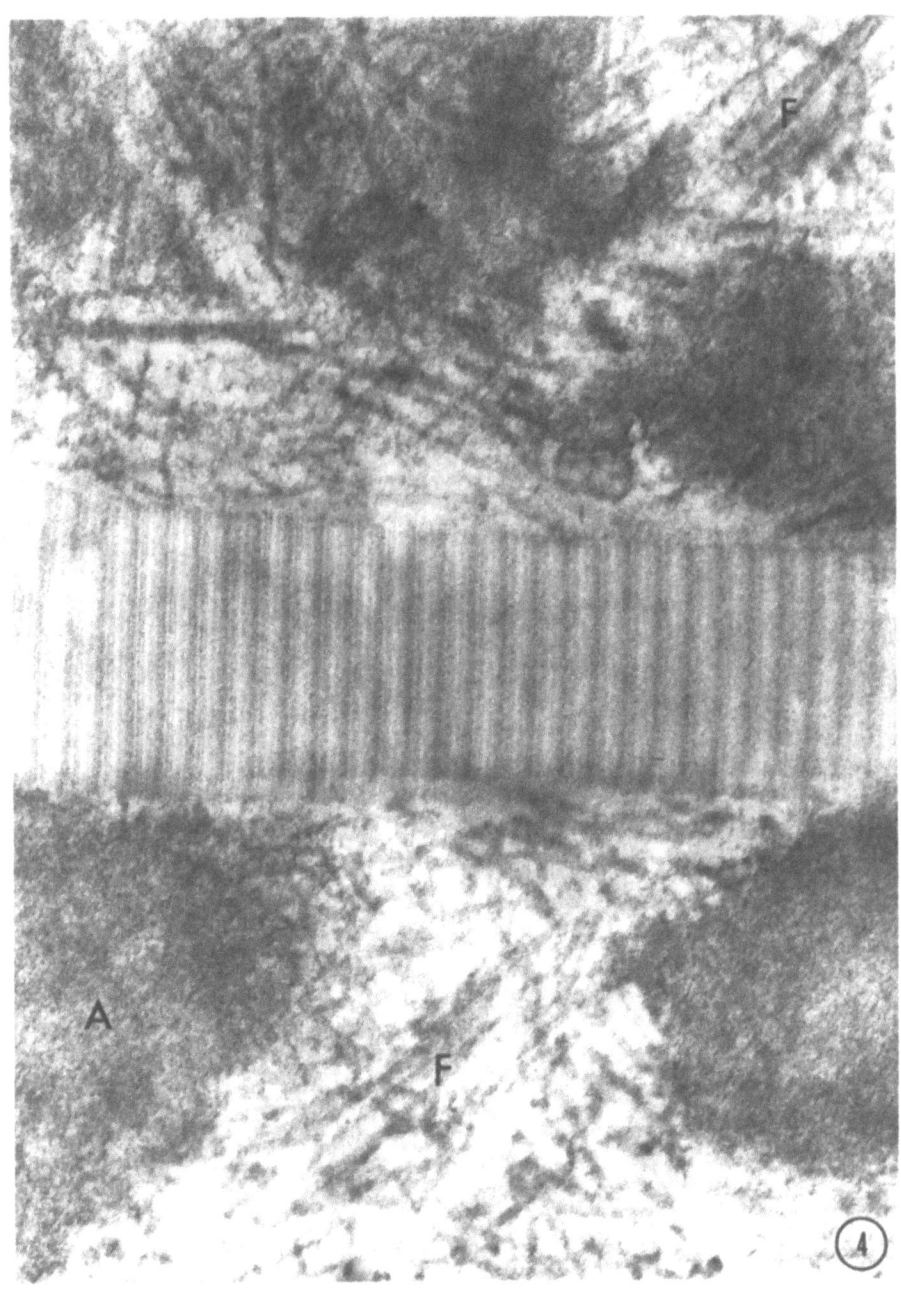

Fig. 4. A higher power (X 75,000) electron micrograph of the extracellular space around notochordal cells to show 426 Å banded collagen fibre, 30–50 Å filamentous matrix material (*F*), and floccular aggregates of electron dense material (*A*). (Courtesy of Dr. G. Strudel)

HH 10 embryos both around the notochord and the ventral spinal cord); (b) 100 Å, tubular, banded fibrils some distance from the notochord, sensitive to removal by hyaluronidase and amylase but not by collagenase. Bazin and Strudel (1972, 1973)

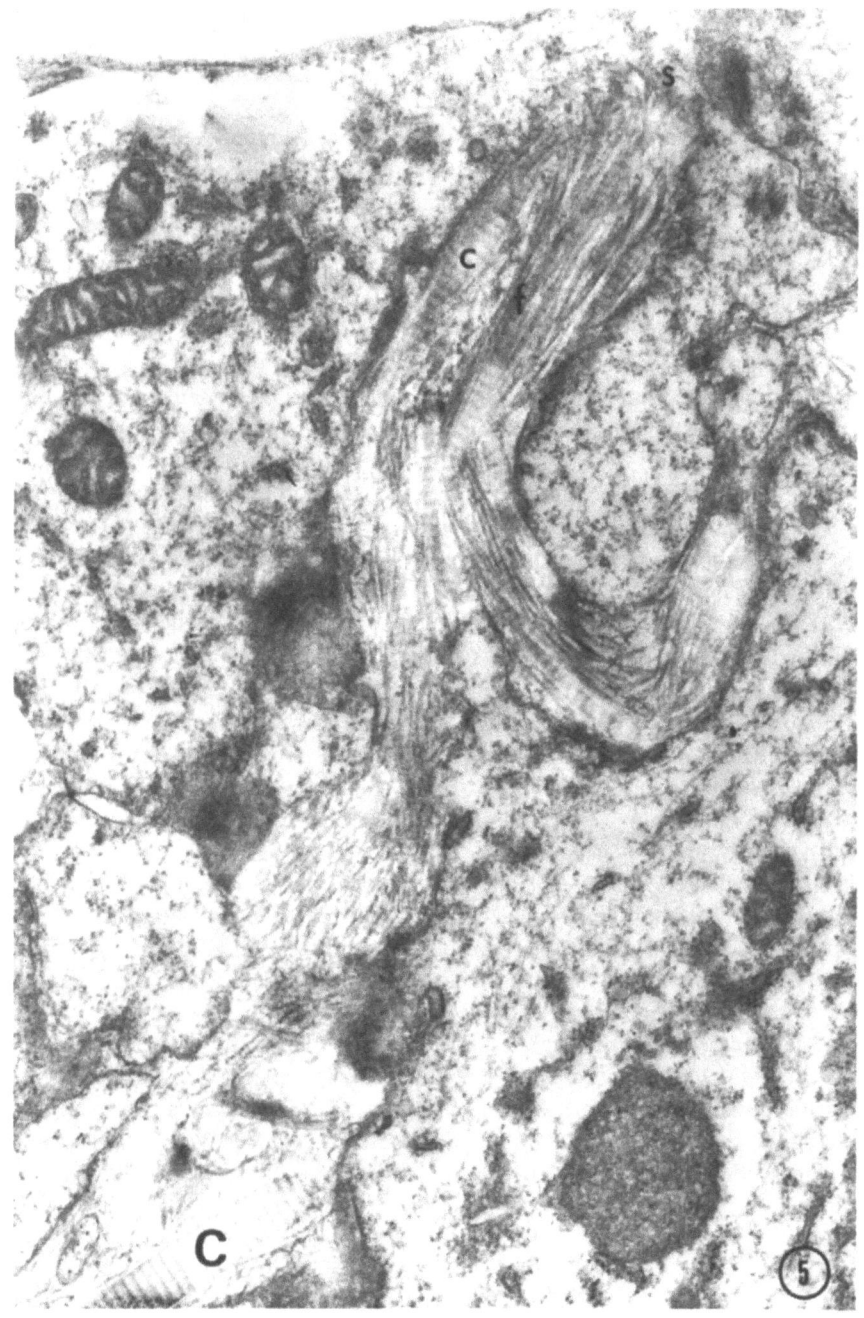

Fig. 5. Details of collagen polymerization in the extracellular space around notochordal cells are shown in this micrograph. Collagen fibres (C) can be seen to be continuous with thinner, banded fibrils (f), which are being extruded from the cell surface (s). X 25,000. (Courtesy of Dr. G. Strudel)

have described collagen fibrils from peri-notochordal and peri-spinal cordal location of 2–3 day-old embryos (Figs. 3–5).

These fibrils are sensitive to collagenase, and in the same position as was uptake of ^{3}H-proline as monitered by autoradiography. They suggested, as did Ruggeri (1972), that this collagenous material became incorporated as part of the extracellular matrix of the differentiating vertebral cartilage.

Is there any direct evidence that the collagen located around the spinal cord is actually produced by the spinal cord? Yes. Cohen and Hay (1971) showed that isolated ventral spinal cord from 2-day-old embryonic chicks synthesized collagen when maintained in isolation in vitro, that the collagen was deposited as fibrils in the extracellular matrix and that it was deposited in association with a basement membrane.

Is there any direct evidence that the collagen located around the notochord is actually produced by the notochord? Yes.

Carlson et al. (1974) isolated the notochord from 2-day old chick embryos by trypsinization and cultured the notochords for 2–3 days. Immediately after isolation, extracellular matrix and basement lamina were lost. After a period in culture, extracellular microfibrils were found adjacent to the newly-constituted basement lamina. Subsequent culturing, in the presence of 100 ug/ml ascorbic acid, showed that the fibrils (some up to 1500 Å wide in sheets) had cross-striations with an axial periodicity of 510 Å – within the range for collagen (Carlson and Upson, 1974).

The collagen from notochord has been partially characterized. Linsenmayer et al. (1973) isolated notochords from $2^1/_2$-day chick embryos in vitro where they were labelled with H^3-proline. The collagen so labelled was shown to comigrate with α 1 (11) collagen isolated from cartilage.

Carboxymethyl cellulose chromatography and cyanogen bromide analysis were carried out by Miller and Mathews (1974) on collagen isolated from the notochord of the sturgeon (*Scaphirhynchus platorhynchus*). The collagen was found to be of the α 1 (11) type with the same amino acid content as sturgeon cartilage collagen, and very similar to avian and mammalian cartilage α 1's.

Trelstad et al. (1973) isolated spinal cords from the posterior region of 2-day-old chick embryos, cultured them for 16 hours and determined (by molecular sieve and ion-exchange chromatography) that the collagen synthesised in vitro was also of the α 1 (11) type. Antibodies against type II collagen bind to the notochord of HH stage 15 embryonic chicks (von der Mark et al., 1976).

The developmental significance, if any, of the fact that tissues (notochord, and spinal cord) which promotes chondrogenesis in sclerotomal cells, produce the same collagen type as does the tissue which they induces (cartilage) has yet to be established. Perhaps rather than saying that notochord possesses cartilage-type collagen we should be saying that cartilage possesses notochord-type collagen.

7. Function of Extracellular Matrix Produced by Notochord and Spinal Cord

This raises the obvious question of whether the collagen and glycosaminoglycans produced by the inducers plays any role in the initiation of chondrogenesis within the somitic mesoderm. This was after-all the motivation for looking at these matrices in the first place. Two experimental approaches have been utilised: — enzymatic removal of extracellular matrix from notochord or spinal cord and testing of retention of inductive ability, or addition of matrix products to otherwise unstimulated somites.

Strudel (1972, 1973 a–c), has cultured axial rudiments from $2^1/_2$ day chick embryos in presence of collagenese (20 μg) or testicular hyaluronidase (10 μg) for 4–5 days and found inhibition of cartilage differentiation within the sclerotome of the explanted rudiment. If such rudiments are transferred to a control medium lacking the enzymes, chondrogenesis commences. Similar treatment of rudiments from 4-day embryos (i. e. after the induction) allowed cartilage differentiation but the cartilage produced was abnormal. The inhibition of cartilage differentiation in the younger rudiments was attributed to the removal of extracellular collagen and glycosaminoglycan (presumably sulphated glycosaminoglycan as testicular hyaluronidase was used) from around the notochord and neural tube. Absence of metachromatic periaxial material after enzyme treatment was confirmed histochemically and ultrastructurally as was the return of such material when rudiments were placed in control media. O'Hare (1972c) put collagenase + hyaluronidase into a millipore filter and then grafted HH 9–12 somites + spinal cord onto the filter. Without enzyme treatment 52 % of the grafts produced cartilage whereas only 14 % of those on enzyme-treated filters did so. He attributed this lack of chondrogenesis to loss of the basement membrane and/or of extracellular matrix from the spinal cord rather than to a direct effect of the enzymes on the somite. Of course, enzymatic treatment of the whole axial rudiment (notchord, spinal cord and somites) removes collagen and glucosaminoglycan from the somites as well as from the other tissues, so that inhibition of chondrogenesis could result from inability of the sclerotomal cells to secrete matrix rather than from inability of the notochord or spinal cord to act inductively. This also applies to studies such as those of Strudel (1975a, b, c). In these studies the proline analogue L-azetidine-carboxylic acid was either injected in ovo (at days zero, one or two) or added to culture media containing somites and axial organs. Secretion of extracellular matrix was retarded and differentiation of the myotome favoured at the expense of the sclerotome.

These problems are overcome by treating isolated notochords with chondroitinase or with testicular hyaluronidase and then combining them with intact somites in an environment which will not allow spontaneous chondrogenesis. Kosher and Lash (1975) have done this and shown that cartilage did indeed not form unless surface extracellular matrix was present on the notochord. In the latter case trypsin-treated notochords (which will not induce) were cultured, perinotochordal material reappeared as did ability to induce. The reformation of the perinotochordal material involves synthesis of new basement membrane, reappearance of micorfibrils, and uptake of ^{3}H proline (Lauscher and Carlson, 1975).

Experiments involving the addition of matrix products to somites have not been as clear cut. Kosher et al. (1973) extracted chondroitin 4 and 6-sulphate from vertebral cartilage of 10 day-old chick embryos or from vertebral cartilage of 13–15 day-old embryos and added it to the medium in which somites (HH 17) were being cultivated. Although the percent of cultures forming cartilage was the same in both treated and control medium, glycosaminoglycan accumulation dropped off after two days in control medium but was maintained at the high initial rate in medium to which extracted proteoglycan had been added. This correlated with increased amount of cartilage in these cultures. The effect was therefore a qualitative, not a quantitative one. Similar results are obtained when procollagen or collagen are added to somites in vitro

– the rate of ongoing chondrogenesis is enhanced, with the type II collagen providing the greatest enhancement (Kosher and Church, 1975). For a comprehensive review of the function of the extracellular matrix see Manasek (1975).

8. Discussion and Conclusions

From the foregoing analysis of past research on chondrogenesis of the somitic mesoderm it is evident that much has been learned of the differentiative abilities of the mesoderm and of the role of tissue-tissue interactions in the evocation of chondrogenesis. The major events occuring in the sclerotome, in the ventral spinal cord and in the notochord during development of the embryonic chick have been summarised in Table 10.

The initial appearance of pre-somitic mesoderm and its subsequent development and growth involves complex cell to cell interactions, the making and breaking of cell junctions between adjacent mesodermal cell, between mesodermal cells and cells of

Table 10. Chronology of vertebral chondrogenesis in the embryonic chick

H. H. Stage	Age	Notochord, spinal cord	Sclerotome
3^+	12 hrs	S^{35} in presumptive notochord	
10	36 hrs	collagen, glycosaminoglycans as extracellular matrix	Intact mass of ovoid cells
11	42 hrs	basement membrane present	Nest of cells breaking up.
12	47 hrs		Stellate cells beginning to migrate
13	50 hrs	youngest age shown to produce collagen when isolated in vitro.	sclerotomites starting to form.
17	60 hrs	considerable hyaluronate around notochord. Treatment with collagenase or hyaluronidase prevents induction.	
18	3 day	notochord vacuolated, now induces 100 % of cultured somites to chondrify.	sclerotomal cells finished migration.
23	4 day		extracellular matrix around cells closest to notochord – now prechondroblasts.
27–30	5–7 day	more chondroitin sulphate than hyaluronate.	extracellular matrix spreading into peripheral cells.
31–36	7–10 day	ventral spinal cord loses inductive ability.	chondroblasts and chondrocytes
36–39	10–13 day	notochord loses inductive ability (HH 37.5).	cell death in area nearest notochord.

the epiblast and between mesodermal cells and cells of the neural plate and notochord. How these cellular events are regulated and how they mediate the early morphogenetic changes in the pre-somitic mesoderm remain unsolved problems.

The mechanism(s) responsible for initial cavitation of the somite and for the differential cellular behaviour of prospective dermatome, myotome and sclerotome also remain unsolved. For example, is hyaluronate or hyaluronidase activity associated with break-up of the sclerotome into individual cells and their migration to a perinotochordal and peri-spinal cordal position? What is the nature of the extracellular space through which they migrate? Do the extracellular matrices of the notochord and spinal cord towards which they are migrating provide any chemotactic guidance or attraction for the migrating cells? On approaching the spinal cord are migrating cells repulsed by the segmentally organizing spinal ganglia, thereby effecting resegmentation of the sclerotome? Techniques of selective ablation of single primordia with follow-up by detailed ultrastructural analysis may be a profitable approach to such questions.

The developmental basis of morphogenesis also remains a major uncharted ocean in the world of somite chondrogenesis. Although Minor (1973) has provided a start in the ultrastructural analysis of regional differences in somite organization along the embryonic axis no comparative stage by stage analyses of the somites is available. In many respects Williams' (1910) paper is still the best available. Whilst such detailed analysis may appear trivial at first sight it is indeed vital for the interpretation of experimental studies in which somites from different levels and ages of embryos are utilized. The state of determination and differentiation attained by the somitic mesoderm, length of time in contact with notochord and spinal cord (and with adjacent epithelia?), even the number of somites co-cultured, all vitally affect the interpretation of experimental results, and have lead, in the past, to diametrically opposed conclusions as to the inherent abilities of the somitic mesoderm.

A further unsolved morphogenetic phenomenon is the recognized dorso-ventral specificity in the inductive ability of the spinal cord. Only the ventral portion of the spinal cord acts inductively upon the sclerotome. The existence of this polarity has been verified experimentally but its basis remains undetermined: — there is some suggestion that only the ventral portion of the spinal cord (in common with the notochord), produces the extracellular matrix which has been shown to play a vital role in the mediation of notochordal and spinal cordal influences to the sclerotomal mesenchyme.

Normal morphogenesis of the neural arches is dependent upon influences from the central nervous system, whilst that of the centrum is dependent upon the notochord. Ablation experiments provide this conclusion but not the underlying mechanism which await investigation.

The segmentally-arranged masses of sclerotomal cells evidently have an inherent potential for chondrogenesis at a basal or minimal rate if isolated in vitro. Depending on the nature of the supplementation of the medium somitic mesoderm from embryonic chicks as young as HH stages 10 (ca. 36 hours of incubation) will chondrify in vitro. Grafting to a vascular environment such as the avian chorio-allantoic membrance will permit somites to chondrify from embryos younger than will chondrify in vitro. The expression of the potential of the sclerotome is extremely sensitive to perturbations of the environment, although how the supplements to the medium exert this influence is unknown. The possibility that they act in the same way as does the extracellular matrix of the notochord and spinal cord bears investigating.

The presence of the synthetic machinery for, and the synthesis of, chondrotin sulphate, abilities previously thought to be conveyed to the pre-somitic mesoderm during the inductive process, are now know to be pathways shared in common with many other non-chondrogenic early embryonic tissues. The unique ability of the somitic mesoderm and of its interacton with notochord and ventral spinal, is its ability to augment the rates of synthesis of chondroitin sulphate and of other glycosaminoglycans and to deposit the synthesised material into an extracellular matrix. The control is not at the level of presence or absence of synthetic pathway but at the level of regulation of activity of a pre-existing pathway. The ability to synthesize glycosaminoglycans is one of the potentials of sclerotomal cells not an event associated with terminal differentiation, nor with transfer of developmental information from the evironment but a pre-existing potential which can be speeded or slowed by environmental enhancement or depletion.

Presence of the notochord and of the ventral portion of the spinal cord are the major in vivo evironmental factors which augment chondrogenesis within the somite. They are also the most potent agents active in vitro. Whilst other factors will substitute for notochord and for spinal cord and allow chondrogenesis to continue in vitro, their absence is not compensated for by the embryo in vivo.

Does this in vivo dependence on notochord and spinal cord justify the term induction? I believe that it does, not because the notochord and spinal cord confer specific developmental information to the somites but because they act as specific in vivo augmenters of existing potentials and as such are necessary for completion of vertebral histogenesis and morphogenesis. Whether the dissipation of spinal cord-induced cartilage is a result of information from the spinal cord and how it acts in vivo remains undetermined.

This "inductive" activity of the notochord and spinal cord is mediated by products of their extracellular matrices, notably collagen and glycosaminoglycans. These are produced by the inducers, remain active after mechanical disruption, are lost after enzymatic treatment but can reform after removal of the enzyme thereby returning inductive ability to notochord and spinal cord.

The epidermal ectoderm may also play in vivo. If it does then it is subordinate to and dependent upon the presence of notochord and or spinal cord. There is sufficient evidence supporting an epithelial involvement in somitic chondrogenesis in vivo to warrent careful experimental evaluation of its role.

Sufficient factors involved in the process of initiation of somitic chondrogenesis have now been studies in vitro that the time is now ripe for a return to the techniques of in vivo manipulation. Do the notochord and spinal cord, so indispensable to the somites in vivo, but replaceable in vitro, act in the embryo in the same manner as they appear to function in the culture dish? How do they enable the sclerotome to regulate synthesis and deposition of glycosaminoglycans and what is the mechanism of their action on the morphogenetic properties of the vertebrae. What of post-chondrogenesis events? Is remodelling of the vertebral arches a response to a post-inductive action on the part of the spinal cord? The studies carried our over the last ten years in vitro have set the stage for a major assault on the control of chondrogenesis in vivo.

9. Summary

The available literature on the development of the somites, their subdivision into sclerotome, myotome and dermatome, and the available knowledge on chondrification of the sclerotomal mesenchyme have been reviewed. Whether the sclerotome chondrifies because of self-differentiation (so called "spontaneous chondrogenesis" in vitro) or because of inductive interaction with the notochord and spinal cord has been discussed by reviewing, chronologically, the grafting experiments of the 1920's and 1930's, the in vivo transplantation and extirpation experiments of the 1940's and 1950's and the in vitro studies of the 1950's, '60's and '70's. The weight of evidence supports a specific inductive role for the notochord and spinal cord whose absence cannot be compensated for in vivo but for which substitutes exist in vitro. The nature of the inductive activity of these tissues was discussed with emphasis on the recent characterization of extracellular matrix products as the inductive agents. Collagen and glycosaminoglycans, both of which are produced by notochord and by ventral portion of spinal cord, can augment the pre-existing bias of the sclerotome for chondrogenesis by increasing the rate at which sclerotome produces cartilage-specific products. How these observations obtained in vitro might relate to in vivo chondrogenesis has been discussed.

Acknowledgements

This paper was begun while the author was Visiting Professor, Department of Biomedical Sciences, University of Guelph, Guelph, Canada. Financial support was provided by the National Research Council of Canada (grant #A−5056 and Travel Grant #T−0688). I have profited from discussions with Professor M. Hardy-Fallding (Guelph), Drs. R. A. Kosher and E. J. Kollar (Univ. Connecticut), and Dr. M. S. Tyler (University of Maine). Dr. G. Strudel's (Director, Ecole Pratique des Hautes Etudes, C. N. R. S., Nogent-sur Marne, France) comments on the manuscript are gratefully acknowledged as are those of Drs. M. S. Tyler and R. A. Kosher.

References

Abbott, J., Mayne, R., Holtzer, H.: Inhibition of cartilage development in organ cultures of chick somites by the thymidine analog, 5-bromo-2'deoxyuridine. Develop. Biol. **28**, 430–441 (1972)

Abrahamson, P. A., Lash, J. A., Kosher, R. A., Minor, R. R.: The ubiquitous occurrence of chondroitin sulfates in chick embryos. J. exp. Zool. **194**, 511–518 (1975)

Amprino, R.: Autoradiographic research on the S^{35}-sulphate metabolism in cartilage and bone differentiation and growth. Acta anat. (Basel) **24**, 121–163 (1955)

Arey, L. B.: Developmental Anatomy. A Textbook and Laboratory Manual of Embryology. Philadelphia: W. B. Saunders, 7th 3d. 1974

Avery, G., Chow, M., Holtzer, H.: An experimental analysis of the development of the spinal column. V. Reactivity of chick somites. J. exp. Zool. **132**, 409–426 (1956)

Balinsky, B. I.: An Introduction to Embryology. Philadelphia: W. B. Saunders 4th ed. 1975

Bancroft, M., Bellairs, R.: The development of the notochord in the chick embryo, studied by scanning and transmission electron microscopy. J. Embryol. exp. Morph. **35**, 383–401 (1976)

Bazin, S., Strudel, G.: Mise en évidence de collagène dans le matérial extracellulaire des organes axiaux de jeunes embryons de Poulet. C. R. Acad. Sci. (Paris) **275**, 1167–1170 (1972)

Bazin, S., Strudel, G.: Biosynthesis of collagen in the axial organs of young chick embryos. In: Biology of Fibroblast (eds. E. Kulonen and J. Pikkarainen) p. 411–416. New York: Academic Press 1973

Bellairs, R.: Developmental Processes in Higher Vertebrates. London: Logos Press 1971

Benoit, J. A. A.: L'otocyste exercé-t-il une action inductrice sur le mésenchyme somatique chez l'embryon de poulet? J. Embryol. exp. Morph. **8**, 39–46 (1960)

Benoit, J. A. A., Schowing, J.: Morphogenesis of the neurocranium. In: Tissue Interactions during Organogenesis. (ed. E. Wolff), p. 105–130. New York: Gordon and Breach 1970

Carlson, E. G., Upson, R. H.: "Native" striated collagen produced by chick notochordal epithelial cells in vitro. Amer. J. Anat. **141**, 441–446 (1974)

Carlson, E. C. Upson, R. H., Evans, D. K.: The production of extracellular connective tissue fibrils by chick notochordal epithelium in vitro. Anat. Rec. **179**, 361–374 (1974)

Chevallier, A.: Rôle du mésoderme somitic dans le développement de la cage thoracique de l'embryon d'oisseau. 1. Origine du segment sternal et méchanismes de la différenciation des côtes. J. Embryol. exp. Morph. **33**, 291–333 (1975)

Cohen, A. M., Hay, E. D.: Secretion of collagen by embryonic neuroepithelium at the time of spinal cord-somite interaction. Develop. Biol. **26**, 578–605 (1971)

Cooper, G. W.: Induction of somite chondrogenesis by cartilage and notochord: A correlation between inductive activity and specific stages of cytodifferentiation. Develop. Biol. **12**, 185–212 (1965)

Corsin, J.: Matériel extracellulaire et chondrogenèse chez les amphibians. Arch. Anat. micr. Morph. exp. **63**, 231–238 (1974)

Crissman, R. S., Low, F. N.: A study of fine structural changes in the cartilage -to-bone transition within the developing chick vertebra. Amer. J. Anat. **140**, 451–470 (1974)

Detwiler, S. R.: An experimental study of spinal nerve segmentation in Amblystoma with reference to the plurisegmental contribution to the brachial plexus. J. exp. Zool. **67**, 395–443 (1934)

Detwiler, S. R.: Observations upon the migration of neural crest cells, and upon the development of the spinal ganglia and vertebral arches in Amblystoma. Amer. J. Anat. **61**, 63–94 (1937)

Detwiler, S. R., Holtzer, H.: The developmental dependence of the vertebral column upon the spinal cord in the urodeles. J. exp. Zool. **132**, 299–310 (1956)

Detwiler, S. R., Dyke, van R. H.: The development and fuctions of deafferented fore limbs in Amblystoma. J. exp. Zool. **68**, 321–346 (1934)

Deuchar, E. M.: Cellular Interactions in Animal Development. London: Chapman and Hall 1975

Ellison, M. L., Ambrose, E. J.: Easty, G. G.: Chondrogenesis in chick embryos somites in vitro. J. Embryol. exp. Morph. **21**, 331–340 (1969)

Ellison, M. L., Lash, J. W.: Environmental enhancement of in vitro chondrogenesis. Develop. Biol. **26**, 486–496 (1971)

Feller, A., Sternberg, H.: Zur Kenntnis der Fehlbildungen der Wirbelkorper bei Spaltbildungen des Zentralnervensystems und ihre Formale Genese. Z. Anat. Entwickl.-Gesch. **103**, 606–633 (1934)

Flowers, M., Grobstein, C.: Interconvertibility of induced morphogenetic responses of mouse embryonic somites to notochord and ventral spinal cord. Develop. Biol. **15**, 193–205 (1967)

Franco-Browder, D., de Rydt, J., Dorfman, A.: The identification of a sulfated mucopolysaccharide in chick embryos stages 11–23. Proc. nat. Acad. Sci. (Wash.) **49**, 643–647 (1963)

Frederickson, R. G., Low, F. N.: The fine structure of perichordal microfibrils in control and enzyme-treated chick embryos. Amer. J. Anat. **130**, 347–376 (1971)

Gearhart, J. D., Mintz, B.: Clonal origins of somites and their muscle derivatives: evidence from allophenic mice. Develop. Biol. **29**, 27–37 (1972)

Glick, M. C., Lash, J. W. Madden, J. W.: Enzymic activities associated with the induction of chondrogenesis in vitro. Biochim. biophys. Acta (Amst.) **83**, 84–92 (1964)

Gordon, J. S., Lash, J. W.: In vitro chondrogenesis and cell viability. Develop. Biol. **36**, 88–104 (1974)

Grobstein, C.: Inductive interaction in the development of the mouse metanephros. J. exp. Zool. **130**, 319–339 (1955)

Grobstein, C., Holtzer, H.: In vitro studies of cartilage induction in mouse somite mesoderm. J. exp. Zool. **128**, 333–357 (1955)

Grobstein, C., Parker, G.: In vitro induction of cartilage in mouse somite mesoderm by embryonic spinal cord. Proc. Soc. exp. Biol. (Med.) **85**, 477–481 (1954)

Hall, B. K.: Correlations between the concentrations of acid mucopolysaccharides and collagen in the tibia of the embryonic chick. Canad. J. Zool. **51**, 771–776 (1973)

Hay, E. D., Meier, S.: Glycosaminoglycan synthesis by embryonic inductors – neural tube, notochord and lens. J. Cell Biol. **62**, 889–898 (1974)

Hoadley, L.: The differentiation of isolated chick primordia in chorio-allantoic grafts. 11. The effect of the presence of the spinal cord, i. e. innervation, on the differentiation of the somitic region. J. exp. Zool. **42**, 143–162 (1925)

Holtfreter, J.: Mesenchyme and epithelia in inductive and morphogenetic processes. In: Epithelial-Mesenchymal Interactions (ed. R. Fleischmajer) p. 1–30. Baltimore: Williams and Wilkins 1968

Holtzer, H.: Morphogenetic influence of the spinal cord on the axial skeleton and musculature. Anat. Rec. **109**, 373–374 (1951)

Holtzer, H.: An experimental analysis of the development of the spinal column. 1. Response of pre cartilage cells to size variations of the spinal cord. J. exp. Zool. **121**, 121–148 (1952a)

Holtzer, H.: An experimental analysis of the development of the spinal column. 11. The dispensability of the notochord. J. exp. Zool. **121**, 573–591 (1952b)

Holtzer, H.: The development of mesodermal axial structures in regeneration and embryogenesis. In: Regeneration in Vertebrates (ed. C. S. Thornton) p. 15–33 Univ. Chicago Press. 1959

Holtzer, H.: Aspects of chondrogenesis and myogenesis. In: Synthesis of Molecular and Cellular Structure (ed. D. Rudnick) p. 335–370. New York: Ronald Press 1961

Holtzer, H.: Control of chondrogenesis in the embryo. Biophys. J. **4**, 239–250 (1964a)

Holtzer, H.: The induction and maintenance of the vertebral cartilages. 2nd Intern. Congress Congenital Malformations p. 233–239. New York: Intern. Med. Congress (1964b)

Holtzer, H.: Induction of chondrogenesis: a concept in quest of mechanisms. In: Epithelial-Mesenchymal Interactions (ed. R. Fleischmajer) p. 152–164. Williams and Wilkins, Baltimore: 1968

Holtzer, H., Abbott, J.: Oscillations of the chondrogenic phenotype in vitro. In: The Stability of the Differentiated State (ed. H. Ursprung) Vol. 1, p. 1–16. Berlin – Heidelberg – New York: Springer 1968

Holtzer, H., Detwiler, S. R.: An experimental analysis of the development of the spinal column. 111. Induction of skeletogenous cells. J. exp. Zool. **123**, 335–366 (1953)

Holtzer, H., Mayne, R.: Experimental morphogenesis. The induction of somitic chondrogenesis by embryonic spinal cord and notochord. In: Pathobiology of Development-or Ontogeny Revisited (eds. E. V. D. Perrin and M. J. Finegold) p. 52–65. Baltimore: Williams and Wilkins, 1973

Holtzer, H., Weintraub, H., Mayne, R., Mochan, B.: The cell cycle, cell lineages and cell differentiation. Current Topics Develop. Biol. 7, 229–256 (1972)

Homes, F. A., van Leeuwen, G., Zilliken, F.: Induction of cell differentiation. 11. The isolation of a chondrogenic factor from embryonic chick spinal cords and notochords. Biochim. biophys. Acta (Amst.) 56, 320–325 (1962)

Horstadius, S.: Ueber die Folgen von Chordaexstirpation an spaeten Gastrulae und Neurulae von Amblystoma punctatum. Acta Zool. (Stockh.) 25, 75–87 (1944)

Johnston, P. M., Comar, C, L.: Autoradiographic studies of the utilization of S^{35}-sulfate by the chick embryos. J. biophys. biochem. Cytol. 3, 231–238 (1957)

Jurand, A.: The development of the notochord in chick embryos. J. Embryol. exp. Morph. 10, 602–621 (1962)

Jurand, A.: Some aspects of the development of the notochord in mouse embryos. J. Embryol. exp. Morph. 32, 1–34 (1974)

Kitchin, I. C.: The effects of notochordectomy in Amblystoma mexicanum. J. exp. Zool. 112, 393–415 (1949)

Kosher, R. A.: Inhibition of "spontaneous", notochord-induced, and collagen-induced in vitro somite chondrogenesis by cyclic AMP derivatives and theophylline. Develop. Biol. 53, 265–276 (1976)

Kosher, R. A., Church, R. L.: Stimulation of in vitro somite chondrogenesis by procollagen and collagen. Nature 258. 327–329 (1975)

Kosher, R. A., Lash, J. W.: Notochordal stimulation of in vitro somite chondrogenesis before and after enyzmatic removal of perinotochordal materials. Develop. Biol. 42, 362–378 (1975)

Kosher, R. A., Lash, J. W., Minor, R. R.: Environmental enhancement of in vitro chondrogenesis. IV. Stimulation of somite chondrogenesis by exogenous chondromucoprotein, Develop. Biol. 35, 210–220 (1973)

Kvist, T. N., Finnegan, C. V.: The distribution of glycosaminoglycans in the axial region of the developing chick embryo. 1. Histochemical analysis. J. exp. Zool. 175, 221–240 (1970a)

Kvist, T. N., Finnegan, C. V.: The distribution of glycosaminoglycans in the axial region of the developing chick embryo. 11. Biochemical analysis. J. exp. Zool. 175, 241–257 (1970b)

Langman, J., Nelson, G. R.: A radioautographic study of the development of the somite in the chick embryo. J. Embryol. exp. Morph. 19, 217–226 (1968)

Lash, J. W.: Tissue interaction and specific metabolic responses: chondrogenic induction and differentiation. In: Cytodifferentiation and Macromolecular Synthesis (ed. M. Locke), p. 235–260. New York: Academic Press 1963a

Lash, J. W.: Studies on the ability of embryonic mesonephros explants to form cartilage. Develop. Biol. 6, 219–232 (1963b)

Lash, J. W.: Normal embryology and teratogenesis. Amer. J. Obstet. Gynecol. 90, 1193–1207 (1964)

Lash, J. W.: Differential behaviour of anterior and posterior embryonic chick somites in vitro. J. exp. Zool. 165, 47–56 (1967)

Lash, J. W.: Phenotypic expression and differentiation: in vitro chondrogenesis. In: The Stability of the Differentiated State (ed. H. Ursprung) Vol. 1, p. 17–24. Berlin – Heidelberg – New York: Springer 1968a

Lash, J. W.: Chondrogenesis: genotypic and phenotypic expression. J. Cell Physiol. 72 (suppl 1), 35–46 (1968b)

Lash, J. W.: Somite mesenchyme and its response to cartilage induction. In: Epithelial-Mesenchymal Interactions (ed. R. Fleischmajer and R. Billingham), p. 165–172. Baltimore: Williams and Wilkins 1968c

Lash, J. W., Glick, M. C., Madden, J. W.: Cartilage induction in vitro and sulfate-activating enzymes. Nat. Cancer Inst. Monogr. 13, 39–49 (1964)

Lash, J. W., Holtzer, S., Holtzer, H.: An experimental analysis of the development of the spinal column. V 1. Aspects of cartilage induction. Exp. Cell Res. 13, 292–303 (1957)

Lash, J. W., Holtzer, H., Whitehouse, M. W.: In vitro studies on chondrogenesis: the uptake of radioactive sulphate during cartilage induction. Develop. Biol. 2, 76–89 (1960)

Lash, J. W., Hommes, F. A., Zilliken, F.: Induction of cell differentiation. 1. The in vitro induction of vertebral cartilage with a low molecular weight tissue component. Biochem. biophys. Acta (Amst) 56, 313–319 (1962)

Lash, J. W., Rosene, K., Minor, R. R., Daniel, J. C., Kosher, R. A.: Environmental enhancement of in vitro chondrogenesis. III. The influence of external potassium ions and chondrogenic differentiation. Develop. Biol. **35**, 370–375 (1973)

Lauscher, C. K., Carlson, E. C.: The development of proline containing extracellular connective tissue fibrils by chick notochordal epithelium in vitro. Anat. Rec. **182**, 151–168 (1975)

Levitt, D., Dorfman, A.: Concepts and mechanisms of cartilage differentiation. Current Topics Develop. Biol. **8**, 103–149 (1974)

Linsenmayer, T. F., Trelstad, R. L., Gross, J.: The collagen of chick embryonic notochord. Biochem. biophys. Res. Commun. **53**, 39–45 (1973)

Lipton, B. H., Jacobson, A. G.: Analysis of normal somite development. Develop. Biol. **38**, 73–90 (1974a)

Liptin, B. H., Jacobson, A. G.: Experimental analysis of the mechanisms of somite morphogenesis. Develop, Biol. **38**, 91–103 (1974b)

Manasek, F. J.: Extracellular matrix-dynamic component of developing embryo. Current Topic-Develop. Biol. **10**, 35–102 (1975)

Marzullo, G., Lash, J. W.: Separation of glycosaminoglycans on thin layers of silica gel. Anal. Biochem. **18**, 575–578 (1967a)

Marzullo, G., Lash, J. W.: Acquisition of the clondrocytic phenotype. Exp. Biol. Med. **1**, 213–219 (1967b)

Mathews, M. B.: Comparative biochemistry of chondroitin sulphate proteins of cartilage and notochord. Biochem. J. **125**, 37–46 (1971)

Mathews, M. B.: Connective Tissue Macromolecular Structure and Function. Molecular Biology, Biochemistry and Biophysics. Vol. 19, p. 1–318. Berlin – Heidelberg – New York: Springer (1975)

Miller, E. J., Mathews, M.: Characterization of notochord collagen as a cartilage-type collagen. Biochem. biophys. Res. Commun. **60**, 424–430 (1974)

Minor, R. R.: Somite chondrogenesis. A structural analysis. J. Cell Biol. **56**, 27–50 (1973)

Mintz, B.: Allophenic mice of multi-embryo origin. In: Methods in Mammalian Embryology. (ed. J. Daniel Jr.), p. 186–214. San Francisco: Freeman 1971

Mintz, B.: Clonal units of gene control in mammalian differentiation. In: Cell Differentiation (eds. R. Harris, P. Allin and D. Viza) p. 267–271. Copenhagen: Munksgaard 1972

Moore, W. J., Mintz, B.: Clonal model of vertebral column and skull development derived from genetically mosaic skeletons of allophenic mice. Develop. Biol. **27**, 55–70 (1972)

Murray, P. D. F., Selby, D. S.: Chorio-allantoic grafts of single somites and of the unsegmental paraxial region of the two-day chick embryo. J. Anat. **67**, 563–572 (1933)

O'Connell, J. J., Low, F. N.: A histochemical and fine structural study of early extracellular connective tissue in the chick embryo. Anat. Rec. **167**, 425–438 (1970)

O'Hare, M. J.: Differentiation of chick embryo somites in chorioallantoic culture. J. Embryol. exp. Morph. **27**, 215–228 (1927a)

O'Hare, M. J.: Chondrigenesis in chick embryo somites grafted with adjacent and heterologous tissues. J. Embryol. exp. Morph. **27**, 229–234 (1972b)

O'Hare, M. J.: Aspects of spinal cord induction of chondrogenesis in chick embryo somites. J. Embryol. exp. Morph. **27**, 235–243 (1972c)

Okayama, M., Pacifici, M., Holtzer, H.: Differences among sulfated proteoglycans synthesized in nonchondrogenic cells, presumptive chondroblasts, and chondroblasts. Proc. Nat. Acad. Sci. (Wash.) **73**, 3224–3228 (1976)

Olson, M. D., Low, F. N.: The fine structure of developing cartilage in the chick embryo. Amer. J. Anat. **131**, 197–216 (1971)

Packard, D. S. Jr., Jacobson, A. G.: The influence of axial structures on chick somite formation. Develop. Biol. **53**, 36–48 (1976)

Patten, B. M., Carlson, B. M.: Foundations of Embryology. 3rd ed. New York: McGraw Hill 1974

Piiper, J.: On the evolution of the vertebral column in birds, illustrated by its development in Larus and Struthio. Phil. Trans. Roy. Soc. (B), **216**, 285–351 (1928)

Pinot, M.: Etude expérimentale de la morphogenèse de la cage thoracique cheq l'embryon de Poulet. Méchanisme et origine du matériel. J. Embryol. exp. Morph. **21**, 149–164 (1969)

Remak, R.: Untersuchungen uber die Entwicklung der Wirbeltthiere. Berlin: G. Reimer, 1855

Ruggeri, A.: Ultrastructural, histochemical and autoradiographic studies on the developing chick notochord. Z. Anat. Entwickl.-Gesch. 138, 20–33 (1972)

Seno, T., Büyüközer, I.: Cartilage formation in somite grafts of chick blastoderms. Proc. Nat. Acad. Sci. (Wash.) 44, 1274–1284 (1958)

Smithberg, M.: The origin and development of the tail of the frog, Rana pipens. J. exp. Zool. 127, 397–425 (1954)

Stockdale, F., Holtzer, H., Lash, J. W.: An experimental analysis of the development of the spinal colum VII. Response of dissociated somite cells. Acta Embryol. Morph. exp. (Palermo) 4, 40–46 (1961)

Straus, W. L. Jr., Rawles, M. E.: An experimental study of the origin of the trunk musculature and ribs in the chick. Amer. J. Anat. 92, 471–509 (1953)

Strudel, G.: Conséquences de l'excision de troncons du tube nerveus sur la morphogenèse de l'embryon de poulet et sur la différenciation de ses organes: Contribution a la genèse de l'orthosympathique. Ann. Sci. Nat. Zool. 15, 251–319 (1953a)

Strudel, G.: Influence morphogenese du tube nerveux et de la chorde sur la differenciation de la colonne vertèbrale. C. R. Soc. Biol. (Paris) 147, 132–133 (1953b)

Strudel, G.: L'action morphogène du tube nerveux et de la corde sur la différenciation des vertèbres et des muscles vertébraux chex l'embryon de poulet. Arch. Anat. micr. Morph. exp. 44, 209–235 (1955)

Strudel, G.: Action inductrice de l'extrait du tube nerveux et de la chorde sur la formation du cartilage vertebral. C. R. acad. Sci (Paris) 249, 470–471 (1959)

Strudel, G.: Induction de cartilage in vitro par l'extrait de tube nerveux et de chorde de l'embryon de poulet. Develop. Biol. 4, 67–86 (1962)

Strudel, G.: Autodifférenciation et induction de cartilage à partir de mésenchyme somitique de poulet chultivé in vitro. J. Embryol. exp. Morph. 11, 399–412 (1963)

Strudel, G.: Some aspects of organogenesis of the chick spinal column. In: Experimental Biology and Medicine. Morphological and Biochemical aspects of Cytodifferentiation. (eds. E. Hagen, W. Wechsler,, and F. Zilliken). Vol. 1, p. 183–198 Basel: S. Karger, 1967

Strudel, G.: Matériel extracellulaire et chondrogenèse vertébrale. C. R. Acad. Sci. (Paris) 272, 473–476 (1971)

Strudel, G.: Differenciation d'ebauches chondrogènes d'embryons de poulet cultivées in vitro sur differents milieux. C. R. Acad. Sci (Paris) 274, 112–115 (1972)

Strudel, G.: Etude de la differenciation du cartilage vertebral. Lyon med. 229, 29–42 (1973a)

Strudel, G.: Relationship between the chick periaxial metachromatic extracellular material and vertebral chondrogenesis. In:Biology of Fibroblast (eds. E. Kulonen and J. Pikkarainen) p. 93–101. New York: Academic Press 1973b

Strudel, G.: Matériel extracellulaire périaxial et chondrogenèse vertébrale. Ann. Biol. 12, 401–416 (1973c)

Strudel, G.: Effect of an L-proline analogue, the L-azetidine-2-carboxylic acid, on the phenotypic differentiation of the chick somitic mesenchyme. C. R. Acad. Sci (Paris) 280, 1007–1010 (1975a)

Strudel, G.: Periaxial extracellular material and vertebral chondrogenesis. In: Protides of the Biological Fluids, 22nd Colloquium, p. 51–58. Oxford: Pergamon Press 1975b

Strudel, G.: Control of the phenotypic vertebral cartilage differentiation by the periaxial extracellular material. In: Extracellular Matrix Influences on Gene Expression (eds. H. C. Slavkin and R. C. Greulich) p. 655–670. New York: Academic Press 1975c

Strudel, G., Gateau, G.: Etude de l'action tératogène du sulfate de nicotine sur les stades jeunes de l'embryon de poulet. C. R. Acad. Sci. (Paris) 272, 2480–2483 (1971)

Sweeney, R. M., Watterson, R. L.: Rib development in chick embryos analyzed by means of tantalum foils blocks. Amer. J. Anat. 126, 127–150 (1969)

Thorp, F. F., Dorfman, A.: Differentiation of connective tissues. Current Topics Develop. Biol. 2, 151–190 (1967)

Toole, B. P.: Hyaluronate turnover during chondrogenesis in the developing chick limb and axial skeleton. Develop. Biol. 29, 321–329 (1972)

Trelstad, R. L., Kang, A. H., Cohen, A. M., Hay, E. D.: Collagen synthesis in vitro by embryonic spinal cord epithelium. Science 179, 295–297 (1973)

Von der Mark, H., von der Mark, K., Gay, S.: Study of differential collagen synthesis during development of the chick embryo by immunofluorescence. 1. Preparation of collagen type 1 and type 11 spefific antibodies and their application to early stages of the chick embryo. Develop. Biol. 48, 237–249 (1976)

Watterson, R. L.: Neural tube extirpation in Fundulus heteroclitus and resultant neural arch defects. Biol. Bull. 103, 310 (1952)

Watterson, R. L., Lowler, I., Fowler, B. J.: The role of the neural tube and notochord in development of the axial skeleton of the chick. Amer. J. Anat. 95, 337–399 (1954)

Williams, J. L.: The development of cervical vertebrae in the chick under normal and experimental conditions. Amer. J. Anat. 71, 153–177 (1942)

Williams, L. W.: The somites of the chick. Amer. J. Anat. 11, 55–100 (1910)

Zilliken, F.: Notochord induced cartilage formation in chick somites: intact tissues versus extracts. In: Experimental Biology and Medicine, Morphological and Biochemical Aspects of Cytodifferentiation. (eds. E. Hagen, W. Wechsler,, and F. Zilliken), Vol. 1, p. 199–212. Basel: S. Karger 1967

Subject Index

Ascorbic acid 36
APS-kinase 30
ATP sulphurylase 30

Basement membrane 8, 28, 32, 36, 37, Table 10
Bone 10, 14, 17, 28
Bromo-2-deoxyuridine (BuDR) 20

Cartilage, as inducer 29, Table 8
Cell death and chondrogenesis 21, Table 10
Cell division and chondrogenesis 21
Centrum 11, 16, 39, Figure 2
Chondroblast 10, 28, Table 10
Chondrocyte 10, 28, Table 10
Chondroitinase 37
Chondroitin sulphate 20, 30–35, 37, 40, Tables 9, 10
Chordomas 11
Clonal cell lines 11
Collagen 10, 30, 31–37, 40, 41, Figures 4, 5 Table 10
Collagenase 33, 34, 36, 37, Table 10
Conditioning of medium 18–21
Connective tissue 9
Costal process 10

Dermatome 8, 9, 23, 24, 39, 41
Dermomyotome 9, 15, 16, 24, Figure 1

Ectoderm 12–14, 17, 25–28, 40, Table 7
Ectopic inducers 25–29, Table 7
Endoderm 12, 13, 17, 25, 27, Table 7
Ependyma 16
Epiblast 8, 39
Epithelia 24, 27, 28, 30, 40, Tables 6, 7
Extracellular matrix of cartilage 10, 29–31, 36, 37, 40
Extracellular matrix of epithelia 28
Extracellular matrix of notochord and spinal cord 8, 10, 13, 29, 31–41, Figures 3–5
Extracellular matrix of sclerotome 10, Table 10

Fibroblast 9, 10, 18

Glucosamine 30

Glycosaminoglycans addition to medium 20, 37
Glycosaminoglycans of cartilage 19, 29–32, 40
Glycosaminoglycans of notochord and spinal cord 10, 30–38, 40, 41, Table 10

Heparin sulphate 32
Hyaluronate 30–32, 38, Tables 9, 10
Hyaluronidase 30, 34, 37, 39, Tables 9, 10
Hypertrophy of cartilage 29, Table 8
Hypertrophy of notochord 22, 23, Table 5

Inducer ectopic 13, 25–29, Table 7
Inducer isolation of, 24, 25
Intervertebral disc 9–11
Irradiation 10, 22, Table 6

L-azetidine carboxylic acid 37

Mesoderm lateral plate 8, 24, 28, Table 6
Mesoderm paraxial 8, 14
Mesoderm pre-somitic 8, 38, 39
Mesoderm somatic 10
Mesoderm somitic 8, 10, 12, 13, 16, 17, 19–21
Migration of neural crest cells 11
Migration of sclerotomal cells 8–11, 14, 32, 39, Figure 1, Table 10
Myoblast 9
Myotome 8–11, 23, 24, 39, 41, Figure 1

Neural arch 11, 14–16, 21, 24, 39, 40 Figure 2
Neural crest cells 11, 16, 17
Neural ectoderm 8, 28
Neural plate 8, 11, 14, 39
Notochord centrum formation and 11, Figure 2
Notochord duration of inductive ability 22, 23, 39, Tables 4, 10
Notochord extracellular matrix of 8, 9, 13, 31–37, 40, 41, Figures 3 and 5, Table 10
Notochord integrity of and induction 21, 22, 25
Notochord mechanical role of 14, 22
Notochord morphogenesis of cartilage and 14, 15, 39, 40

Notochord segmentation of mesoderm and 8, 39
Notochord surgical removal of 13–16, Figure 2
Nucleus pulposus 11

Osteoid 10
Otic capsule 28, Table 7

Primitive streak 8, 31

Resegmentation of sclerotome 10, 11
Ribs 10, 16

Sclerotome 8–10, 14–16, 23, 27, 28, 31, 32, 36, 37, 39–41, Figure 1, Table 10
Sclerotomite 10, 11, Table 10
Segmentation of paraxial mesoderm into somites 8, 14
Segmentation of sclerotome 11, 15, 39
Self differentiation of cartilage 12–20, 41
Somite effect of mass of 13, 14, 16–21, 39, Table 2
Somite organizing center for 8
Somite segmentation of 8, 9, 39, 41, Figure 1

Somite sequence of chondrification in 9, 18–20
Spinal cord dorsal portion of 17, 39
Spinal cord duration of induction 22, 23, 39, Tables 4 and 10
Spinal cord extracellular matrix of 8, 10, 13, 31–41, Table 10
Spinal cord integrity of, and induction 21, 22, 24, 25
Spinal cord morphogenesis of cartilage and 15, 16, 39, 40
Spinal cord surgical removal of 13, 15, 16
Spinal cord ventral portion of 9, 10, 15–17, 32, 36, 39–41, Table 10
Spinal ganglia 8, 11, 12, 16, 39, Figure 2
Spontaneous chondrogenesis 17–19, 37, Table 1

Tetraparental mice 11

UDP-N-acetylglucosamine 30

Vacuolation of notochord 23, Tables 5 and 10
Ventral spinal cord and induction 9, 10, 15–17, 32, 36, 39–41, Table 10

Author Index

Abbott, J. 11, 20
Abrahamsohn, P. A. 30
Amprino, R. 30
Arey, L. B. 9, 10
Avery, G. 13, 17, Tables 1 and 7

Balinsky, B. I. 11
Bancroft, M. 23, 32
Bazin, S. 34
Bellairs, R. 8
Benoit, J. A. A. 28, Table 7

Carlson, E. C. 36
Chevallier, A. 10
Cohen, A. M. 36
Cooper, G. W. 17, 18, 22–24, 29, Tables 5 and 8
Corsin, J. 32
Crissman, R. S. 10

Detwiler, S. R. 11, 13, 15, 21
Deuchar, E. M. 8

Ellison, M. L. 18, 20, Table 2

Feller, A. 11
Flowers, M. 17
Franco-Browder, D. 30, 31
Frederickson, R. G. 37

Gearhart, J. D. 11
Glick, M. C. 30
Gordon, J. S. 19, 21
Grobstein, C. 13, 16, 17, 22, 28, Table 7

Hall, B. K. 31
Hay, E. D. 32
Hoadley, L. 13
Holtfreter, J. 15
Holtzer, H. 7, 11, 13, 15, 17–21, 24, 25, 27, Table 7
Hommes, F. A. 25
Horstadius, S. 13, 14

Johnston, P. M. 30, 31
Jurand, A. 32

Kitchin, I. C 14
Kosher, R. A. 19, 20, 37, 38
Kvist, T. N. 31

Langman, J. 9
Lash, J. 7, 17–20, 22, 25, 27, 30, 31 Table 7
Lauscher, C. K. 37
Levitt, D. 7
Linsenmayer, T. F. 36
Lipton, B. H. 8

Manasek, F. J. 38
Mark von der, H. 36
Marzullo, G. 30
Mathews, M. B. 32
Miller, E. J. 36
Minor, R. R. 9, 21, 32, 33, 39
Mintz, B. 11
Moore, W. J. 11
Murray, P. D. F. 13, 16

O'Connell, J. J. 31, 32
O'Hare, M. J. 19, 22–24, 27, 28, 37, Tables 6 and 7
Okayama, M. 30
Olson, M. D. 10

Packard, D. S. Jr. 8
Patten, B. M. 10
Piiper, J. 10
Pinot, M. 10

Remak, R. 10
Ruggeri, A. 36

Seno, T. 27, Table 7
Smithberg, M. 15
Stockdale, F. 17, 27, Table 7
Straus, W. L. 14
Strudel, G. 7, 10, 11, 12, 16, 18, 21, 22, 24, 25, 27, 28, 32, 37, Table 7
Sweeney, R. M. 10

Thorp, F. F. 18, 25
Toole, B. P. 30, Table 9
Trelstad, R. L. 36
Tremaine, R. 22

Watterson, R. L. 13, 15, 16, 21
Williams, J. L. 11, 14
Williams, L. W. 8, 19, 39

Zilliken, F. 18, 25

Other Reviews of Interest in this Series

Volume 48

Part 1: **Böck, P.**: Das Glomus caroticum der Maus. 49 figures. 84 pages. 1973. ISBN 3-540-06368-4

Part 2: **Sousa-Pinto, A.**: Cortical Projections of the Medial Geniculate Body in the Cat. 19 figures. 42 pages. 1973. ISBN 3-540-06477-X

Part 3: **Vanpeperstraete, F.**: The Cartilaginous Skeleton of the Bronchial Tree. 42 figures. 80 pages. 1973. ISBN 3-540-06536-9

Part 4: **Oksche, A; Farner, D. S.**: Neurohistological Studies of the Hypothalamo-Hypophysial System of Zonotrichia leucophrys gambelii (Aves, Passeriformes). With Special Attention to its Role in the Control of Reproduction. 74 figures. 136 pages. 1974. ISBN 3-540-06586-5

Part 5: **Scheuermann, D. W.**: Über den Feinbau des Myocards von Rana Temporaria (L) Ultrastructure of ventricular cardiac muscle of Rana temporaria. 31 figures. 70 pages. 1974. ISBN 3-540-06609-8

Part 6: **Reinboth, R.; Simon, N.**: Adenohypophyse und Hypothalamus. Histophysiologische Untersuchungen bei Lepomis (Centrarchidae) 41 figures. 85 pages. 1974. ISBN 3-540-06749-3

Volume 49

Edinger, T.: Paleoneurology 1804-1966, an Annotated Bibliography. 258 pages. 1975. ISBN 3-540-07060-5

Volume 50

Part 1: **Aldskogius, H.**: Indirect and Direct Wallerian Degeneration in the Intramedullary Root Fibres of the Hypoglossal Nerve. An Electron Microscopical Study in the Kitten. 59 figures. 78 pages. 1974. ISBN 3-540-06750-7

Part 2: **Vigh-Teichmann, I.; Vigh, B.**: The Infundibular Cerebrospinal-Fluid Contacting Neurons. 24 figures. 91 pages. 1974. ISBN 3-540-06979-8

Part 3: **Raedler, A.; Sievers, J.**: The Development of the Visual System of the Albino Rat. 16 figures. 88 pages. 1975. ISBN 3-540-07079-6

Part 4: **Ribi, W. A.**: The Neurons of the First Optic Ganglion of the Bee (Apis mellifera). 21 figures. 43 pages. 1975. ISBN 3-540-07096-6

Part 5: **Halata, Z.**: The Mechanoreceptors of the Mammalian Skin. Ultrastructure and Morphological Classification. 11 figures. 77 pages. 1975. ISBN 3-540-07097-4

Part 6: **Beckers, H. W.; Eisenacher, W.**: Zur Morphologie der Papilla fungiformis einiger Primaten und des Menschen. Zur Morphologie der Papilla fungiformis einiger Nagetiere. Rasterelektronenmikroskopische, licht- und elektronenmikroskopische Untersuchungen. 27 figures. 117 pages. 1975. ISBN 3-540-07098-2

Volume 51

Part 1: **Putte, S. C. J. van der**: The Development of the Lymphatic System in Man. 33 figures. 60 pages. 1975. ISBN 3-540-07204-7

Part 2: **Raedler, A., Sievers, J.**: Influences of Experimental Brain Edema on the Development of the Visual System. 27 figures. 60 pages. 1975. ISBN 3-540-07205-5

Part 3: **Pexieder, T.**: Cell Death in the Morphogenesis and Teratogenesis of the Heart. 52 figures. 100 pages. 1975. ISBN 3-540-07270-5

Part 4: **Svendgaard, N. A.; Björklund, A.; Stenevi, U.**: Regnerative Properties of Central Monoamine Neurons. 24 figures. 77 pages. 1975. ISBN 3-540-07299-3

Part 5: **Gossrau, R.**: Die Lysosomen des Darmepithels. 74 figures. 95 pages. 1975. ISBN 3-540-07271-3

Part 6: **Thorn, L.**: Die Entwicklung des Cortischen Organs beim Meerschweinchen. 23 figures. 97 pages. 1975. ISBN 3-540-07301-9

Volume 52

Part 1: **Ibrahim, M. Z. M.**: Glycogen and its Related Enzymes of Metabolism in the Central Nervous System. 13 figures. 89 pages. 1975. ISBN 3-540-07454-6

Part 2: **Cau, P.; Michel-Bechet, M.; Fayet, G.**: Morphogenesis of Thyroid Follicles in Vitro. 16 figures. 66 pages. 1976. ISBN 3-540-07654-9

Part 3: **Tiedemann, K.**: The Mesonephros of Cat and Sheep. Comparative Morphological and Histochemical Studies. 47 figures. 119 pages. 1976. ISBN 3-540-07779-0

Part 4: **Haug, F.-M. S.**: Sulphide Silver Pattern and Cytoarchitectonics of Parahippocampal Areas in the Rat. Special Reference to the Subdivision of Area Entorhinalis (Area 28) and its Demarcation from the Pyriform Cortex. 49 figures. 73 pages. 1976. ISBN 3-540-07850-9

Part 5: **Phillips, I. R.**: The Embryology of the Common Marmoset (Callithrix jacchus). 22 figures. 47 pages. 1976. ISBN 3-540-07955-6

Part 6: **Nobiling, G.**: Die Biomechanik des Kieferapparates beim Stierkopfhai. 25 figures. 52 pages. 1977. ISBN 3-540-08038-4

Volume 53

Part 1: **Baur, R.**: Morphometry of the Placental Exchange Area. 37 figures. 65 pages. 1977. ISBN 3-540-08159-3

Part 2: **Kaufmann, P.; Davidoff, M.**: The Guinea-Pig Placenta. 21 figures. 91 pages. 1977. ISBN 3-540-08179-8

Part 3: **Hadžiselimović, F.**: Cryptorchidism. Ultrastructure of Normal and Cryptorchid Testis Development. 43 figures. 72 pages. 1977. ISBN 3-540-08361-8

Springer-Verlag Berlin Heidelberg New York

MIX
Papier aus verantwortungsvollen Quellen
Paper from responsible sources
FSC® C105338

If you have any concerns about our products,
you can contact us on
ProductSafety@springernature.com

In case Publisher is established outside the EU,
the EU authorized representative is:
**Springer Nature Customer Service Center GmbH
Europaplatz 3, 69115 Heidelberg, Germany**

Printed by Libri Plureos GmbH
in Hamburg, Germany